HYDRAULIC
TROUBLESHOOTER

Published 2007 by arima publishing

www.arimapublishing.com

ISBN 978-1-84549-192-5

Printed and bound in the United Kingdom

Typeset in Garamond

Abramis is an imprint of arima publishing

arima publishing
ASK House, Northgate Avenue
Bury St Edmunds, Suffolk IP32 6BB
t: (+44) 01284 700321

www.arimapublishing.com

CONTENTS

1

FOREWORD

Power Hydraulics are used in many industries on a world wide basis from steel mills to yachts including use in Mobile, Agricultural, Industrial, Municipal, Automotive and Marine industries from very small power levels to very large. There are many machines that would not exist in their present form without the use of hydraulic power.

The aim of this book is to provide 'users' of hydraulics an understanding of how the various components in a system work. 'Users' means anyone who is involved in designing, developing, using and maintaining hydraulic systems.

The book is titled "Troubleshooter". As well as providing an understanding of how things work it will hopefully provide enough information to enable anyone faced with a hydraulic problem to solve it and keep their systems in good working order.

The information in this book is presented by the Authors in good faith but they accept no responsibility for any consequences arising from its use. The component illustrations in the book are intended only to represent the typical product not the full range available in each type.

2

BACKGROUND

Don Seddon's first interest in hydraulics was when he was an apprentice in 1956-61 at Cincinnati Milling Machines Ltd where hydraulic systems were widely used to drive machine tool tables and headstocks including servo systems for die sinking and copy milling. At that time the first Numerically Controlled Machine tools were being developed.

After a period with Vickers Systems as an Applications Engineer and Denison Deri as Area Sales Manager he was appointed Chief Engineer at Abex Denison in 1968. At that time Abex Denison was a major employer in the UK with 250 personnel at the Burgess Hill facility.

In 1973 he started A&D Fluid Power Ltd the basic product being a range of valves based on the new ISO size 3 mounting surface. The company moved to Havant in 1978 and expanded the range of products to include control valves and circuit blocks mainly for the Marine, Mobile and Agricultural Industries.

The company was taken over by Smiths Industries in 1992 and the Author was retained by them as a Consultant Valve and System Designer until 1999. Two international patents were granted for valve designs during this period.

Sailing has been an interest for the Author over the last 25 years and this interest together with an insight into how engines work in boats led to his first book *Diesel Troubleshooter* being published in 1996. This book has sold over 10,000 copies and is now published in German, Spanish and English. It is considered a valuable reference book for boat owners of all types.
REF ISBN 1 898660 81 8 John Wiley & Sons.

Nick Peppiatt of Hallite Seals International Ltd. who has contributed the chapter on Seals has many years experience in the hydraulic industry and has made a major contribution to the understanding of seals in fluid power applications and the adoption of standards for seals.

If any reader has anything to add to this book or would like to see changes made please contact the Author through the publishers.

FORMAL QUALIFICATIONS

Don Seddon   C Eng MI Mech E
Nick Peppiatt   BSc, PhD, C Eng MI Mech E

BFPA   British Fluid Power Association has been involved with and published many standards and guidelines for fluid power equipment. A full list is available at the back of this book.

NFPA   National Fluid Power Association has done similar work in the USA

In the USA the Society of Automotive Engineers (SAE) has a wide range of standards for hydraulic components.

ISO International Standards Organisation in consultation with standards organisations worldwide has provided many standards for fluid power equipment. A full list is available at the back of this book.

Both Authors have been members of BFPA, BSI, and ISO technical committees for many years.

Principles of Hydraulic System Design by Peter J Chapple was published in 2003 by Coxmoor Publishing Co ISBN 1901892 15 8 and is available from BFPA.

The author thanks the companies who have provided the illustrations and tables reproduced in the book and his family for the help given in preparing it.

3

HOW IT WORKS

In hydraulic systems power is transmitted by pumping hydraulic fluid into a system which has a resistance. The resistance is the pressure at which the system works. This is determined by the load.

The flow rate from the pump determines the speed of movement of the load.

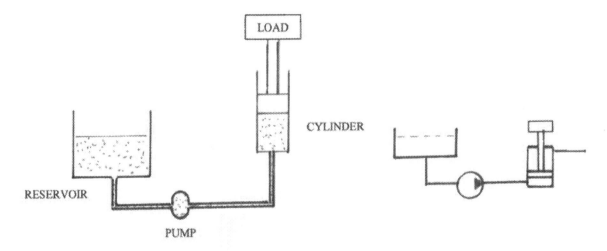

Illustration 1. Basic circuit

In the illustration the pressure required to move the load equals the load divided by the area of the cylinder.

Pressure $= \dfrac{\text{Load}}{\text{Area}}$

The speed of movement of the load is the flow rate from the pump divided by the area of the cylinder.

$$\text{Velocity} = \frac{\text{Flow Rate}}{\text{Cylinder Area}}$$

Hydraulic power depends on flow and pressure.

$$\text{i.e.} \quad \text{Power} = \frac{\text{Pressure x Flow}}{\text{Constant}}$$

The constant depends on the units used.

$$\text{Hydraulic Power KW} = \frac{\text{Pressure (Bar) x Flow (L/Min)}}{600}$$

$$\text{Hydraulic Power HP} = \frac{\text{Pressure (psi) x Flow (USg/m)}}{1715}$$

$$1\text{HP} = 0.7457 \text{ KW}$$

For convenience of calculations and presentation of data the commonly used units for pressure are "bar" and "psi". In some areas of engineering the Pascal (Pa) is used, this is a pure unit but relatively clumsy to use in power hydraulics as one Pa equals one Newton per square metre. To convert to "bar" divide by 10^5.

Flow rates are normally in litres/min, ml/min, US galls/min and cubic inches/min

See Chapter 24 for conversion factors.

EFFICIENCY

All of the various elements in a system absorb some power. This is in the form of friction or pressure drop and internal leakage.

For instance the pump input power is more than the output the difference being a combination of leakage from outlet port to inlet port and mechanical losses due to friction.

Flow of fluid through pipes results in pressure drop due to friction i.e. resistance to flow.
A cylinder will not have any leakage but will have some friction to overcome to start it and keep it moving.

A hydraulic motor will have a small amount of port to port leakage and friction.

Control valves contribute losses due to pressure drop and small amounts of internal leakage.

Well designed pumps, pipework, actuators and control valves will keep these losses to a minimum but nevertheless they cannot be eliminated.

How important these losses are depends on the type of machine but in all cases power is expensive to produce and wastage needs to be kept to a minimum.

The use of very accurate computer controlled machines in the manufacturing of components has resulted in significant gains in efficiency of hydraulic equipment in recent years.

LIFE: Design life of hydraulic components is many thousands of hours and with good maintenance and good system design the hydraulic components can be expected to outlast the life of the machines to which they are fitted (with the possible exception of flexible hoses).

FATIGUE: All mechanical devices subject to fluctuating or cyclic loads have a life expectancy before failure. Hydraulic components are no exception.

GOOD PRACTICE: Use components within their design limitations, keep systems clean and leakfree, inspect regularly. Avoid generation of pressure spikes and load induced uncontrolled pressures.

INTERNAL LEAKAGE

Leakage can take place past valve spools, pump gears, rotors, pistons, motor rotating parts etc.

Leakage through a thin slot i.e. spool clearance is proportional to t^3 and inversely proportional to the viscosity, t being the radial clearance.

Therefore doubling the clearance results in 2^3 i.e. 8 times the leakage. This is why clearances are maintained to very small values of just a few micron.

1 micron = $\dfrac{1}{1000}$ mm = $\dfrac{1}{25400}$ inches = 0.00004 inches (0.0001inch = 2.54 micron)

Typical clearances are: Directional valve spool to body 2.5 to 5 micron
Piston to bore in piston pumps 2.5 to 20 micron
Gear to side plate in gear pump 0.5 to 5 micron
Vane tip to cam ring in vane pump 0.5 to 1 micron

Leakage past a piston or spool is assumed to be laminar flow and this formula can be used to calculate the theoretical flow rate in ml/min.

$$Q = \dfrac{0.00057\ (P_1\text{-}P_2)\ t^3 w}{v\ l}$$

Where (P_1-P_2)   = Pressure difference in bar
    t   = Radial clearance (thickness of slot) in microns
    w   = Width of slot in mm (circumference for round components)
    v   = Viscosity in cSt
    l   = Length of flow path in mm
    Q   = Flow rate in ml per minute

COMPRESSION AND DECOMPRESSION

When a system is subject to pressure there is a small amount of compression of the fluid due to minute particles of entrained air and a small amount of expansion of all the elements in the system that are enclosing the pressure. A rule of thumb guide is that the fluid volume in the system will increase by 0.5% per 70Bar(1000p.s.i.).

Where flexible hoses are used this will be higher.

When decompressing i.e. changing direction or removing the pressure the compressed volume expands rapidly causing shock in the system. A rule of thumb is that when the increased volume exceeds 100cm³ (6in³) then measures should be taken to allow the system to decompress under control. Measures to control decompression should always be considered at the design stage of a system.

4

FLUIDS

The widespread use of hydraulic systems became a reality early in the 20th century when mineral oil was introduced as an operating fluid. The original water hydraulic systems were used extensively in large equipment, dockside cranes, dock gates, elevators etc. Today water is again being used as a fluid in many applications where oil is not acceptable but in the majority of applications mineral oil is used.

Basically this is a lubricating oil which is adapted for hydraulic systems by controlling the viscosity within predetermined limits. The quality of the oil is carefully controlled in terms of water content, oxidisation, contamination and filterability. Additives are included to stop foaming and to improve wear rates of pumps and motors.

The use of mineral oil enables system control components to be manufactured from ferrous materials, cast iron, steel etc., at reasonable cost and to a very high standard. Nitrile rubber compatible with mineral oil provides a durable material for seals (see Seals chapter).

To select a fluid for a particular system the initial criteria are based on the requirements of the pump. The fluid must be within the parameters laid down by the pump manufacturer. They will have carried out extensive tests to determine the range of viscosity that the pumps will function in and the additives required.

The valves, cylinders, motors etc. in the system will generally function with any fluid that is compatible with the pump.

Factors affecting the pump include type of pump, speed of rotation, operating pressure, inlet conditions and range of temperatures over which the pump is expected to operate.

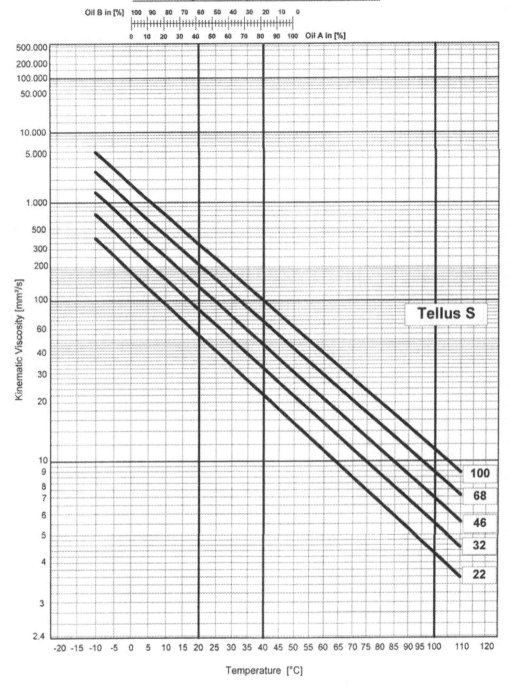

Illustration 2. Viscosity/temperature graph for a range of hydraulic fluids (Shell)

A typical operating viscosity specification for a piston type pump could be 20 to 60 centistokes, with a start up viscosity of 220 centistokes. There are of course conflicting requirements i.e. fluid needs to be viscous (thick) to keep component wear and leakage to a minimum but thin to allow the pump to suck in the fluid without cavitation (see pump section) and flow easily through the piping. It is crucial to use the correct fluid. Failure to do so can result in failure of the pump.

Pressure affects the viscosity, the viscosity doubling over the range zero to 350 bar

Hydraulic oils are classified in terms of viscosity according to ISO 3448 with the number being the viscosity in centistokes at 40°C and atmospheric pressure.

i.e.   ISOVG 32 has a viscosity of 32cSt at 40°C
         ISOVG 68 has a viscosity of 68cSt at 40°C

         (VG = Viscosity Grade)

1 cSt = 10 to the power 6 metres squared per second = 1mm squared per second
Kinematic viscosity (cSt) is the Dynamic viscosity divided by the density.

To keep the oil viscosity within the pump limits it may be necessary to employ heaters and/or coolers in the system.

Mineral oils tend to degrade more quickly over 60°C.(140°F). Systems developed in temperate climates will inevitable run hotter in a tropical environment. It may be necessary to specify a fluid with a higher viscosity to cater for this temperature increase. The reverse applies in Arctic climates.

Oil companies and filter manufacturers offer facilities to report on the condition and contamination of fluid samples.

WHAT CAN GO WRONG

Mineral oil can be contaminated with water which will result in reduced pump and motor life. If the appearance of the fluid is cloudy or milky then water may be mixed in with it.

Always store oil drums on their sides in cool, dry locations so that water does not collect around the outlet and penetrate into the drum.

Zinc antiwear additives can degrade silver. Check with the fluid supplier if using components incorporating silver plated parts.

Contamination from the system and the atmosphere may be present. Effective filtration is needed to keep the fluid to a standard compatible with the system. New oil as delivered may not meet the requirements of the system.

If water is present in the bottom of a reservoir it may be colonised by micro-organisms which live in the water and feed off the oil. The effluent from these organisms forms a jelly like substance which can block filters and also cause corrosion around the bottom of the tank. If this is a persistent problem treatment with a biocide may provide a cure; refer to fluid supplier for advice. Biocides are of course toxic to humans and a great deal of care is required when using them to avoid human contact with the treated fluid.

A simple test for the presence of water in the oil is to heat a small sample in a metal dish. If bubbles rise from the point where the heat is applied and there is a crackling noise water is present.

FIRE RESISTANT FLUIDS

In some industries - coal mining, metal production, die casting etc. where hydraulic fluid spillage would promote fires, fluids which have fire resistant qualities have been developed. These fall into three general types:-

> Oil and water emulsions
> Water/Glycol combination
> Synthetic fluids

Use of these fluids in equipment designed for mineral oil generally results in lower limitations of performance than for mineral oil and nitrile seal materials may not be compatible particularly with synthetic fluids. There are some seal materials which are more compatible such as fluoroelastomer and silicone which can be used.

Non-compatibility results in seals swelling or shrinking, hardening or softening and generally results in a very short life with consequent external leakage. (See Chapter 15).

Flexible hoses also contain rubber or plastic which will be affected in the same way as seals with failure as a result.

WHEN TO CHANGE THE FLUID

In general the harder the fluid works the more frequently it needs to be changed.

Static machinery applications with large reservoirs operating well within the pump limits at temperatures below 60°C(140°F) and with good filtration etc. may only need changing every year.

For mobile machinery working at the limits under all types of weather conditions with small reservoirs, every 500 hours may be appropriate with a different viscosity range being used in winter and summer. Hydraulic oils degrade more rapidly at temperatures over 60°C (140°F).

SAFETY

When handling oils be aware that it is good practice to protect the skin i.e. use gloves. Wash carefully if contact with the skin is experienced.

Dispose of used fluid at an authorised collection point. Do not contaminate drains, soil or water.

5

PIPEWORK

Pipework used in hydraulic systems varies depending on the type of industry. For instance the steel industry keeps pipework to an absolute minimum and uses steel pipes with connecting flanges welded onto the pipes. Control valves are incorporated into or onto circuit or manifold blocks.

In mobile, agricultural, automotive and marine applications the pipework is usually a combination of steel and flexible pipes using compression fitting and pre-formed pipework.

The biggest source of confusion is the wide selection of thread systems used for connecting pipes to components. In Europe the "R" threads based on a Whitworth British Standard Pipe (BSP) parallel thread is widely used as are metric threads generally in conjunction with a bonded steel/rubber washer to provide a seal. For hydraulic applications the thread tolerances are tighter than for water plumbing components.

Other methods of sealing include a copper washer, and a step in the face of the male part which bites into the spotface of the component body.

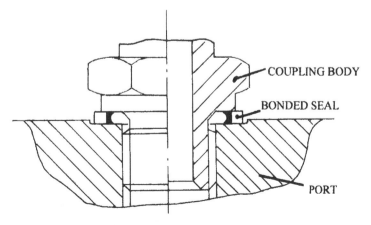

Illustration 3. 'R' threaded connector with bonded seal

TAPER PIPE THREADS

In the USA, NPT or American Standard Taper Pipe Threads are used extensively in industrial applications. In theory they are used without a sealant but in practice a proprietary thread sealant or adhesive applied to the external thread is used to ensure a leaktight connection.

IMPORTANT NOTE: Taper thread fittings can induce heavy bursting forces in component bodies when tightened in addition to and in excess of the forces generated by the internal pressure.

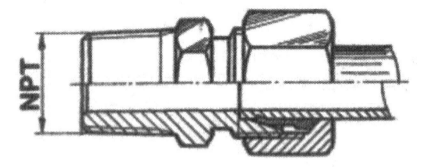

Illustration 4. NPT taper thread connector (Schwer)

S A E THREADS

Also used in mobile, agricultural, marine and automotive systems in the USA and worldwide is the SAE fitting based on UNF threads. The fitting incorporates a chamfer on the female thread and an undercut adjacent to the head of the male part. A rubber "O" ring is fitted into the annular space provided and compressed to provide a seal when the joint is tightened.

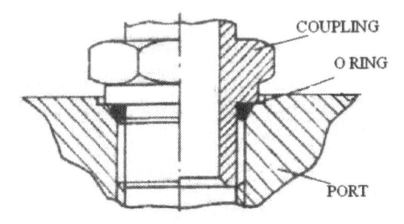

Illustration 5. SAE connector

In Europe a similar metric thread and seal system is widely used.

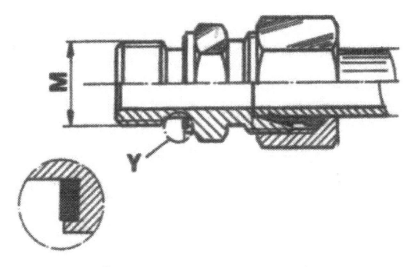

Illustration 6. Metric connector (Schwer)

Heavy duty mobile machinery uses a 2 or 4 bolt steel flange incorporating a seal fitted in a recess which seals against the face of the component. There is an American SAE standard for these items. Some component manufacturers provide flanges made specifically to fit their equipment.

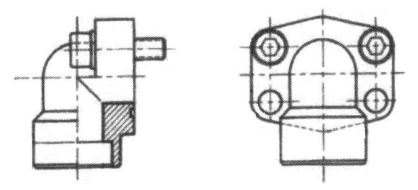

llustration 7. SAE flange connector (JBJ)

For connecting the fitting to the pipe and for pipe to pipe connections compression fittings are widely used.

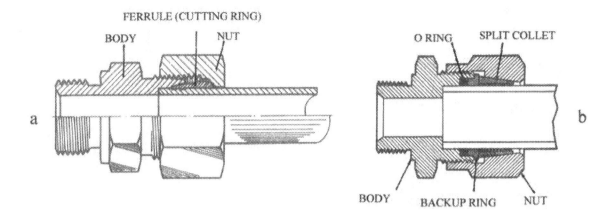

Illustration 8. Compression fittings (a) with ferrule (b) with collet and 'O' ring

When making up or replacing compression fittings it is essential to make them up in a vice rather than on the machine to ensure that the pipes are square with the fitting and the correct bite is obtained with the ferrule.

Pipes must be cut square and de-burred and cleaned thoroughly of all debris, swarf etc. Any metallic debris can cause system component failure. Obtain the manufacturers instructions for the fittings being used. They will advise the specific methods and torque required.

Pipes must have adequate support at least every 1.5 metres for pressure lines, 3 metres for return lines.

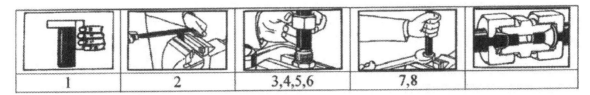

Illustration 9. Assembling a compression fitting (Betabite)

1. Cut the tube to length and file ends square
2. Remove internal and external burrs from tube end and clean thoroughly
3. Make the joint whilst the coupling body is held firmly in a bench vice
4. Lubricate the parts especially the internal body cone, the rear of the ferrule and the internal thread of the nut. On stainless steel couplings the use of a copper based lubricant is required to avoid 'galling'.
5. Slide the nut onto the tube, followed by the ferrule. The open end of the nut should be towards the end of the tube and similarly the cutting or smaller end of the ferrule should point towards the tube end.
6. Present the tube, nut and ferrule to the coupling body making sure that the tube passes cleanly through the nut and ferrule and butts firmly against the step (abutment face) provided in the coupling body. Screw the nut onto the coupling body until finger tight
7. Hold the tube in one hand and with a spanner in the other hand, tighten the nut until the ferrule is felt to just grip the tube. This point is determined by rotation or slightly rocking the tube. From this point the nut should be tightened in accordance with the manufacturer's recommendation
8. If the nut is now removed, the ferrule will have cut its own seating on the tube and whilst it will be found to rotate, it cannot be moved towards the tube end. The joint may now be re-assembled by retightening the nut as per the manufacturer's recommendations. The above procedure will ensure a safe and successful joint.
9. Copper based lubricants are available from fitting manufacturers.

WHAT CAN GO WRONG

Pipes and fittings are subject to high parting forces due to hydraulic pressure. For example the parting force on a 25mm dia pipe fitting at 210 bar is

$$\text{Parting Force} = \text{Pressure times (x) Area of pipe section}$$

$$= 210 \times \frac{\pi d^2}{4} \qquad\qquad 1 \text{ bar} = \frac{10N}{cm^2}$$

$$= \frac{210 \times 10 \times \pi \times 2.5^2}{4} \qquad (25mm = 2.5cm)$$

$$= 210 \times 10 \times \pi \times \frac{6.25}{4}$$

$$= 10308 \text{ Newtons}$$

To obtain force in lbs divide by 4.448
To obtain force in kgs divide by 9.81

or alternatively

$$\text{parting force lbs} = 3045 \text{ psi} \times \pi \times \frac{0.984^2}{4}$$

$$= 2315 \text{ lbs}$$

$$\text{where } 1 \text{ bar} = 14.5 \text{ psi}$$

To avoid failure it is essential to follow the manufacturers recommendations for making up fittings. Use high quality threads and use torque wrenches to apply the correct torque where threaded fittings are fixed to system components. Every fastening or fitting used in a hydraulic system should have a recommended tightening torque.

If pipes and fittings are used underwater careful selection of materials such as stainless steel or alum-nickel-sil-brass is essential and protection by means of sacrificial anodes must be used to avoid crevice corrosion and degradation due to electrolysis.

FLEXIBLE HOSES

For pressure carrying hoses the body of the hose consists of layers of rubber or plastic and wire or fabric reinforcement cut to length then fitted with end connections. The end connections can be straight or angled and are fixed to the hose by swaging an external sleeve onto the hose which is supported by an internal sleeve.

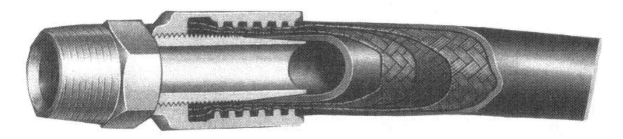

Illustration 10. Typical hose end connection (Parker)

The use of cone type end connections allows the hose to be connected without twisting. Hoses can also be made up with compression fitting ends or flanges to connect onto solid pipework.

The action of a hose in flexing has an effect on the life expectancy of the hose and all manufacturers recommend a minimum bend radius and an arrangement that avoids the hose twisting or stretching.

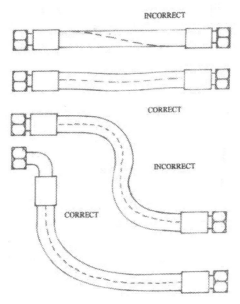

Illustration 11. Hose installation recommendations

Use swivel nuts on ends to allow hose twist to correct itself when being fitted.
Never create a bend smaller than the manufacturer's recommended minimum radius.
Do not allow the hose to twist when operating.
Keep well within the manufacturer's maximum pressure limits.

Whilst flexible hoses are fairly flexible when there is no pressure in them, they are anything but flexible when pressurised. There are considerable forces acting on them to straighten out. This must be borne in mind when determining how they are positioned and restrained. In some instances it may be beneficial to consider rotary joints instead.

The maximum pressure at which a hose can be used is always marked on the hose. Bear in mind that pressure spikes caused by shock loading of the actuator can result in higher pressures that the theoretical system pressure. In load control system cranes etc. the pressure at the cylinder can be higher than the rest of the system due to load induced pressure. Pulsating output from hydraulic pumps can lead to fatigue.

As a rule of thumb use hoses rated at twice the value of the maximum system pressure.

WHAT CAN GO WRONG

Twisting and small bend radiuses will inevitable lead to short hose life. The parting forces for end fittings due to pressure are high and end fittings should be fixed at the correct torque.

Pressure generated by heat in trapped volumes of fluid can be enough to damage pipes and fittings (See Chapter 19 Temperature).

External abrasion will reduce the wall thickness of hoses and lead to eventual failure.

Nitrile hoses suitable for hydraulic oil do not like being immersed in water for long periods.

Hoses and associated pipework can be vulnerable to mechanical damage caused by other machinery working in conjunction with the host machine.

The most vulnerable fluid power components' hoses should be inspected regularly and critical system protected by a hose burst valve or automatic brake. Hose failure can result in a machine failing or moving out of control.

There is a bewildering array of threads, hose sizes, end adaptors and fittings. Using the wrong part for a repair will inevitable result in hose failure.

BFPA statement D14 A simple rule for re-ending hydraulic hoses – **don't.**
BFPA statement D15 Hose assemblies – **don't mix and match.**

Where hoses are used a maintenance programme needs to be established and must include checking:-

a) Hose slippage relative to end fitting
b) Damage to cover
c) Hard, heat cracked or charred hose
d) Leaks at end fittings or from the hose
e) Kinked, crushed or twisted hose
f) Blistered, loose cover or otherwise degraded hose.

Protective sleeves are available to slip over hoses to provide some protection.

HOSE BURST VALVES

Sometimes called a velocity fuse they are positioned between the actuator and the hose. The function of the valve is to close completely should the flow through it exceed the normal system flow i.e. a high flow caused by an actuator running out of control due to a burst hose.

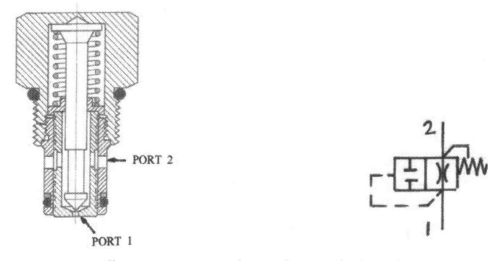

Illustration 12a. Hose burst valve cartridge (Eaton)

Port 1 is connected to the cylinder. The outer poppet slides up and cuts off the flow when the flow setting is reached.

The function of this valve can sometimes produce unintended consequences especially if more than one actuator is involved.

This is a compromise and another solution may be to use a load control valve between the actuator and hose. (See Load Control Valves).

QUICK RELEASE COUPLINGS

Used extensively in agricultural applications where the pipework for various implements is attached to the tractor by means of QR couplings. Also used in vehicles, off-road construction equipment etc.

The coupling consists of two halves male and female. Both are attached to pipework. One is fixed, one flexible by means of a threaded pipe connection.

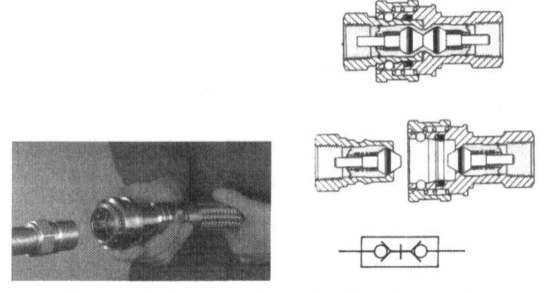

Illustration 12b. Hansen quick release coupling (Guyson International)

The connection is made by drawing back the locking sleeve on the female part and inserting the male part past the seal and then pushing it home. This opens the spring loaded cone valves which are incorporated in each half. Letting go the sleeve at the same time locks the two parts together.

WHAT CAN GO WRONG

Care is needed to ensure dirt does not accumulate on them when not in use. Always use protective covers.

Mechanical damage can occur which can cause leakage past the seal.

If the cone valve tip on the male part is worn down the cone valve will not open fully when connected and cause a high pressure drop in the coupling.

The flow passages through the couplings are fairly complex and the pressure drop through them needs to be considered when selecting them.   Seal failure can occur due to contamination.

The materials used in the coupling need to be good quality to obtain a long life.

Disconnecting the couplings under pressure is not recommended.   Connecting them under pressure is not possible due to the forces involved.

PIPE AND HOSE SIZES

Fluid flowing through a pipe encounters some resistance (pressure drop) so the pipe should normally be sized to keep the resistance to a minimum. In some systems however, pressure drop in pipework is not critical and higher flow velocities can be tolerated than in systems where it is vital to keep the pressure drops to a minimum.

The viscosity of the fluid has an effect on the pressure drop. For turbulent flow, a tenfold increase in viscosity results in an 80% increase in pressure drop as pressure drop is proportional to $v^{0.25}$.

The velocity of fluid through a pipe is the flow divided by the area.

$$\text{i.e.} \quad \frac{\text{Flow ml/sec}}{\text{Area cm}^2} \qquad\qquad 1\text{ml} = 1\text{cm}^3$$

Flow in litres/min needs to be converted to ml/sec i.e. multiplied by $\frac{1000}{60}$

and the flow area of the pipe needs to be in cm^2

$$\text{area} = \frac{\pi d^2}{4} \quad \text{where d the tube inside diameter is in cm}$$

The velocity will then be in cm/sec which can be converted to the more usable term of m/sec by dividing by 100 or feet/sec by dividing by 30.5.

Low pressure pipes and hoses for the inlet side of pumps or for fluid transfer need to be relatively larger to keep pressure drops to an absolute minimum particularly when used as the inlet pipe of pumps working in a open circuit i.e. pump being supplied directly with fluid from a reservoir.

<u>Fluid velocity parameters</u>

Inlet lines	0.5 to 1.5 metres/sec
Pressure lines	2 to 6 metres/sec
Return lines	1.5 to 3 metres/sec

Pressure Loss in Pipes

Flow (l/min)	Tube bore size (mm)								
	5	7	10	13	16	21	25	30	36
1	0.69	0.22							
2	1.38	0.44							
3	2.07	0.66	0.17						
5	4.14	1.24	0.24						
7.5	6.55	1.72	0.31						
10		3.10	0.38	0.14					
15		5.38	0.69	0.21	0.08				
20			1.10	0.30	0.14				
30			2.21	0.69	0.25	0.04			
40				1.17	0.45	0.08	0.04		
50					0.59	0.12	0.07	0.03	
75					1.13	0.23	0.14	0.06	0.02
100						0.41	0.22	0.13	0.03
150							0.45	0.23	0.06
200								0.41	0.10
250									0.16

This chart gives the approximate pressure drop in smooth bore straight pipes, in bar per 3 m length. Bends and fittings will increase the above pressure losses and manufacturers should be consulted for more accurate figures.

Illustration 12c Pressure loss in pipes.(BFPA)

Use low velocities in low pressure or continuously running systems.

When replacing pipework whether fixed or flexible scrupulous cleanliness is required and pipe must be thoroughly cleaned by mechanical means and flushing before being installed.

Under no circumstances weld pipes in situ as weld debris will inevitably stick to the inside of the pipe, break loose and damage the pump.

It is tempting to use PTFE tape for sealing taper thread joints - don't. Any shreds of tape created by screwing the joint together or remaking it in the future will inevitably jam a valve or other critical component in the system.

Try and avoid taper thread joints where the seal is made by tightening the male thread into the female. If there is no alternative thoroughly clean the parts and use a proprietary liquid thread adhesive or sealant applied to the male thread to seal the joint.

6

RESERVOIRS

INDUSTRIAL AND MARINE

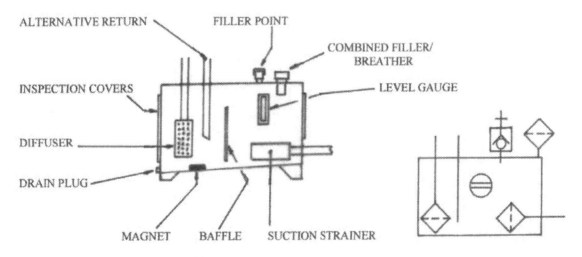

Illustration 13. Industrial reservoir

POSITION RELATIVE TO PUMPS

In open circuits the pumps draw fluid direct from the reservoir and rely on the atmosphere pressure to provide fluid flow into the pump. If the reservoir is positioned above the pump there is some assistance from the height difference in the order of 0.5 psi per foot or 0.1 bar per metre. Where the reservoir is below the pumps the pressure is negative and it may be necessary to reduce the pump speed to obtain satisfactory filling of the pump without cavitation.

FILLER BREATHER

Provides a strainer for fluid being introduced and to filter air being drawn in as the level changes due to the displacement of fluid in the cylinders. This filter should provide filtration of the air moving in and out of the reservoir to the same standard as the fluid filters used in the system.

Ideally new fluid should be introduced to a reservoir through a filter.

Illustration 14. Traditional filler breather (Parker)

LEVEL GAUGE

Visible from the outside of the reservoir they sometimes include a temperature gauge. In automotive power packs this is often a dipstick attached to the filler cap.

HIGH LEVEL

When all cylinders in the system are fully retracted the fluid in the reservoir will be at its highest level. This is when the fluid level should be topped up if necessary.

LOW LEVEL

When the cylinders in the system are fully extended the fluid volume absorbed by the cylinders is equal to rod area times stroke. In the case of single acting cylinders it is by the cylinder area times stroke.

The reservoir design must take this displacement of fluid into account and have sufficient fluid left in for the system to function properly when the cylinders are extended.

The volume of fluid required is the full displacement of single acting cylinders and the rod displacement for double acting cylinders.

Example: Fluid displacement of two double acting cylinders with 100 mm (10cm) diameter rods and a stroke of 1 metre (100cm) would be

$$\frac{2 \times 10^2 \; \pi \times 100}{4}$$

$$= \frac{2 \times 10000 \times \pi}{4}$$

$$= 15200 \text{ cm}^3$$

$$= 15200 \text{ ml}$$

$$= 15.2 \text{ litres} \qquad \text{Divide by 3.8 to obtain US galls}$$

To dissipate heat and maintain fluid quality a general rule is to have a volume of 2 to 4 times the pump capacity per minute in the reservoir at the low level.

SUCTION STRAINER

Normally constructed from wire mesh with a mesh size of 120 micron they provide nominal protection for the system i.e. keeping out large debris. They must be adequately sized to ensure minimum pressure drop and positioned as low as possible with approx 25 mm clearance at the bottom of reservoir.

DIFFUSER

Cylindrical construction with many holes around the periphery the objective is to diffuse the effect of a high velocity jet of oil coming from the return pipe and causing turbulence and foaming.

It is used mainly when baffles are not installed which serve the same purpose. The function of baffles is to circulate fluid around the reservoir so that the same oil is not being used continuously and the reservoir is a more effective cooling medium. They also allow debris flushed through the system into the reservoir to settle on the bottom.

Magnets are sometimes included to collect ferrous debris.

On many occasions reservoirs are used to dissipate heat. They should not be obstructed and airflow around them should be encouraged. For this reason feet are often fitted to raise the reservoir off the floor.

All reservoirs must have inspection covers either on the top or side to clean out the reservoir as part of a regular service schedule or after a component failure. To facilitate this a drain valve needs to be used.

MOBILE AGRICULTURAL AND AUTOMOTIVE RESERVOIRS

These have the same function as industrial reservoirs but generally are smaller and are often incorporated into a machine member. The same principles of fluid flow airflow etc. apply.

In many automotive systems the pump is situated in the reservoir e.g. tailgate lifts, bogie lifts, cab tilt i.e. the reservoir is part of a complete power pack.

The size of these reservoirs is normally determined by the displacement of the cylinders in the system. There must be sufficient volume above the minimum level to allow for the full extension and retraction of the cylinders. (See previous calculations for low level).

RESERVOIR MATERIALS

Welded steel is used mostly and the inside is left untreated. In some applications it is advisable to finish the inside to avoid rusting. This finish must be compatible with the fluid. For the smaller automotive applications moulded plastic reservoirs are often used.

7

HEAT EXCHANGERS (COOLERS)

All wasted energy in hydraulic systems results in heat generated. In most systems the temperature builds up to a working level and stays there whilst the system is functioning.

In some systems a heat exchanger is required to keep the fluid temperature to a working level.

Systems used in mobile, agricultural or automotive machinery generally use air coolers i.e. radiators incorporating fans which are driven either hydraulically, mechanically or electrically.

Sizing the coolers is at best a complicated calculation since there is some heat loss from system components. Heat exchanger manufacturers can offer specialised advice regarding the application of their products.

Industrial and marine systems generally use heat exchangers which use water as the cooling medium. Cold water flows through a number of tubes arranged in parallel immersed in a container through which hydraulic fluid flows in the opposite direction. The cooler must be piped in such a way that it is kept full of water and hydraulic fluid at all times.

If high fluid pressures are expected which may potentially cause tube/seal failure use the cooler in a separate circuit with its own pump.

Coolers using sea water must have internal and external parts which are compatible.

It is good practice to ensure that the hydraulic fluid pressure is higher than the water pressure so that should a leak occur the hydraulic fluid will not be contaminated.

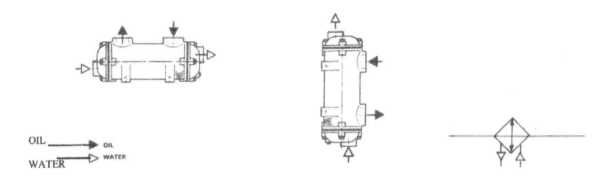

Illustration 15. Water based cooler (Bowman)

The oil cooler should be mounted as shown and piped for counter flow.

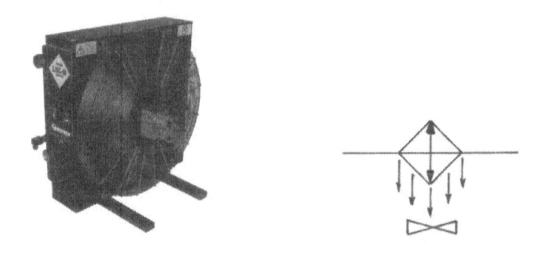

Illustration 16. Air based cooler (Oiltech)

All systems function best within a temperature range.

Too low a temperature results in inefficiency due to the high fluid viscosity too high a temperature puts the function of pumps, motors and sealing components at risk. It can also cause excessive oxidisation of the fluid.

The fluid viscosity must be within the recommendations for the pump being used.

Heat exchangers which work well in temperate climates are less effective in tropical climates and at high altitudes. If a system on a vessel, for instance, has to be used worldwide then advice should be obtained when specifying the coolers. This of course also applies to IC engines including generator sets. Warmer water and air temperatures make cooling more of a requirement and paradoxically more difficult to achieve.

Fire resistant fluids generally have lower working temperature tolerances due to evaporation etc. and if any doubt exists seek advice from the fluid suppliers.

8

FILTERS

Fluid power components, valves, pumps, motors etc. incorporate very accurately machined components and operate with clearances of just a few microns.

1 micron = 1μm = 0.001mm   0.0001 inch = 2.54 micron

To obtain long service life it is essential to incorporate good quality filtration into the systems. Contamination of fluid causes component failure. This also applies to the air filter which cleans the air passing in and out of the reservoir. The air filter should be to the same standard as the fluid filters.

Typical hydraulic component clearances

Component	Microns
Anti-friction bearing	0.5
Vane pump (vane tip to cam ring)	0.5 — 1.0
Gear pump (gear to side plate)	0.5 — 5.0
Servo valves (spool to sleeve)	1.0 — 4.0
Directional valves (spool to body)	2.0 — 5.0
Hydrostatic bearings	1.0 — 25
Piston pump (piston to bore)	5.0 — 40
Servo valve flapper wall	18 — 63
Actuators	50 — 250
Servo valve orifice	130 — 450

Illustration 17. Typical clearances in components (Parker)

When introducing fluid into the reservoir initially, it should be pumped in through a filter and whenever the level is topped up the same procedure should be followed.

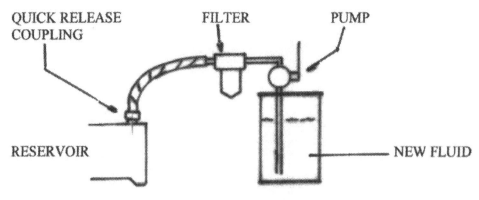

Illustration 18. Filling reservoir

Assuming that the fluid in the reservoir is acceptably clean then a return line filter can be used to filter fluid being returned from the system to keep it clean. A certain amount of debris will be flushed out of the system during the first few hours of running despite the measures taken by the component manufacturers, pipe fitters etc. to keep it to a minimum and over a period of time some wear will take place resulting in metallic debris. A return line filter will collect this.

For sophisticated systems using very high pressure/servo valves/proportional valves etc. pressure line filters between the pump and the system may be a necessity.

Most systems incorporate a suction strainer in the reservoir, this is mainly to keep larger debris out of the pump and offer little resistance to the flow of fluid into the pump. (See Reservoir Chapter).

Clogging indicators measure the pressure drop across the filter element and give an indication of when the element should be changed or cleaned. They are incorporated in the pressure line and return line filters and can include sensors to provide a signal to control stations. They often indicate the first signs of component failure in a system.

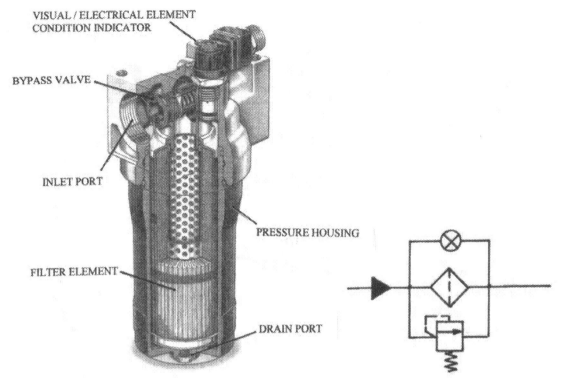

VISUAL / ELECTRICAL ELEMENT
CONDITION INDICATOR

BYPASS VALVE

INLET PORT

FILTER ELEMENT

PRESSURE HOUSING

DRAIN PORT

Illustration 19. Typical filter construction (Parker)

BYPASS VALVES

Bypass valves built into the filter open when the pressure drop across the element is well below its maximum failure parameter and bypass fluid around the element to avoid a collapse or failure.

This protects the filter but it allows some fluid to bypass the element reducing the effectiveness of the filter. Whilst it is desirable to protect the element ideally the filter should also incorporate a clogging indicator which will give a warning before the bypass pressure is reached.

For critical applications where a bypass valve is not acceptable i.e. servo valve systems etc. the element may have a higher pressure rating.

Closed loop systems use a small pump to make up the leakage and provide a constant boost pressure. A pressure line filter is fitted between the boost pump and the system.

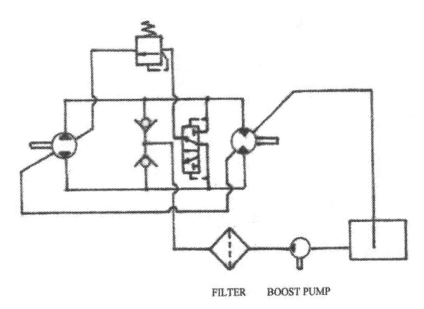

FILTER   BOOST PUMP

Illustration 20. Boost pump and filter for closed loop transmission (simplified circuit)

FILTER ELEMENTS

Filter elements can be made from different materials; cellulose, wire mesh, fibreglass etc. The material controls the effectiveness and efficiency of the filter. For holding capacity and filter efficiency refer to your filter supplier.

Fluid cleanliness is measurable and can be compared to an ISO code number. The grade which is required depends on the components used in the system. A typical range of codes would be as per the table below.

Fluid Cleanliness Required for Typical Hydraulic Components

Components	ISO Code
Servo control valves	16/14/11
Proportional valves	17/15/12
Vane & Piston pumps / motors	18/16/13
Directional & pressure control valves	18/16/13
Gear pumps / motors	19/17/13
Flow control valves, cylinders	20/18/15
New unused Fluid	20/18/15

Illustration 21. Fluid cleanliness typical components (Parker)

The code indicates the maximum number of particles in 1 millilitre (1cc) of fluid. The first part indicates the maximum number greater than 4 micron. The second part indicates the maximum number greater than 6 micron and the third part is the maximum number greater than 14 micron. The actual numbers for instance for 18/16/13 would be:-

18 = 4µm+ = 1300-2500
16 = 6µm+ = 320-640
13 = 14µm+ = 40-80

Filter manufacturers and oil companies offer facilities to test samples of fluid for contamination and oil characteristics. Samples and analysis kits can also be obtained for "in house" monitoring of fluid samples. Test points in the form of self sealing couplings can be fitted at various points in the circuit to enable samples to be drawn off for analysis.

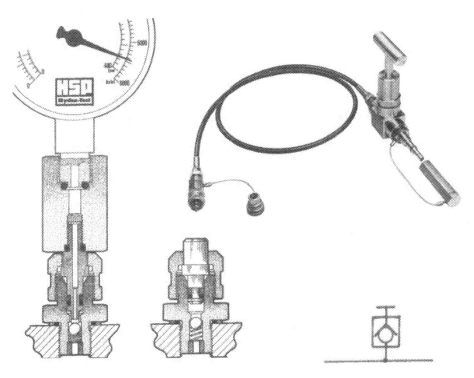

Illustration 22. Test points (HSP)

FILTERS AND PRESSURE

The author's experience indicates that "the higher the system pressure the better the filtration required". Whilst difficult to quantify as it is only one element to be considered, I would list the filtration requirements as follows:

0 – 70 bar	basic
70 - 140 bar	good
140 - 210 bar	to component supplier's standards
210 bar +	better than component supplier's standards

9

ACCUMULATORS

As the name suggests accumulators store hydraulic energy and provide it at appropriate times.

Accumulators were widely used in water hydraulic systems in the industrial revolution, and they consisted of heavy weights bearing down on pistons. The pressure generated was the weight divided by the area of the piston and the volume the area of the piston times the stroke. The biggest accumulators of all are dams which provide pressure and flow for turbines to generate electrical power.

Modern accumulators are either a cylinder with free floating piston with pressurized gas (Nitrogen) on one side of the piston and the system fluid on the other or a cylinder incorporating a flexible membrane with pressurized gas on one side and the system fluid on the other.

As gas is compressed and expanded the pressure generated will change and the precharge pressure has to take account of these changes. Precharge pressure is the gas pressure when the fluid is fully discharged.

SAFETY

It is necessary to connect the accumulator to the system through an isolator with facilities for shutting off the accumulator and draining the pressure from it when the system is stopped and for maintenance and shutdown purposes. Accumulator discharge valve assemblies are available from valve suppliers. They can automatically discharge the accumulator when the system is stopped. Stored energy is very dangerous when released accidentally. It can do considerable damage in an instant.

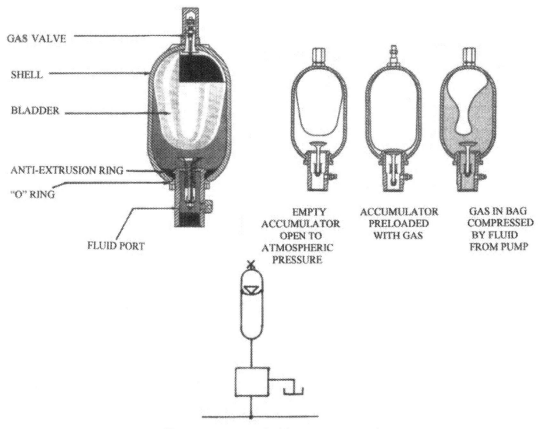

GAS VALVE

SHELL

BLADDER

ANTI-EXTRUSION RING

"O" RING

FLUID PORT

EMPTY
ACCUMULATOR
OPEN TO
ATMOSPHERIC
PRESSURE

ACCUMULATOR
PRELOADED
WITH GAS

GAS IN BAG
COMPRESSED
BY FLUID
FROM PUMP

Illustration 23. Bladder type accumulator

Accumulators should be treated with the utmost respect and only be serviced by qualified personnel. However, in certain systems, there are advantages in being able to obtain high flow rates for short periods of time which makes the selection of accumulators appropriate.

Small accumulators are often used in circuits to absorb pressure spikes and provide a measure of pressure stability.

10

PUMPS

The simplest pumps are operated by hand and these incorporate a piston, lever and two non return valves.

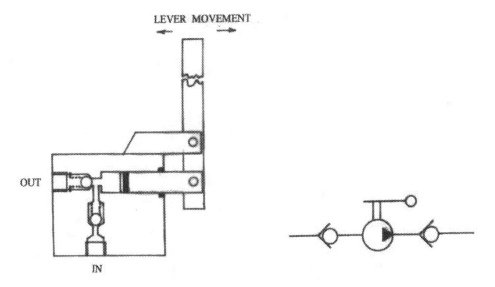

Illustration 24. Hand operated pump (diagrammatic section)

Hand operated pumps are incorporated in many systems as a safety measure i.e. in the event of power failure and in low power systems are used as the direct source of power i.e. on yachts to tension rigging, raising and lowering centreboards etc.

POWER DRIVEN PUMPS

Power driven pumps come in many different forms. This chapter describes the most common forms and their characteristics.

GOOD PRACTICE

For all power driven continuously running pumps arrange the circuit so that the pump starts and stops offload or at reduced pressure by means of open centre directional valves, vented relief valves etc.

AIR DRIVEN PUMPS

Used to provide a source of hydraulic power from a pneumatic supply, these pumps use a reciprocating pneumatic cylinder to drive a hydraulic pump.

Widely used for low flow high pressure applications such as machine tool clamping etc, and where safety demands that electrics cannot be used.

EXTERNAL GEAR PUMPS

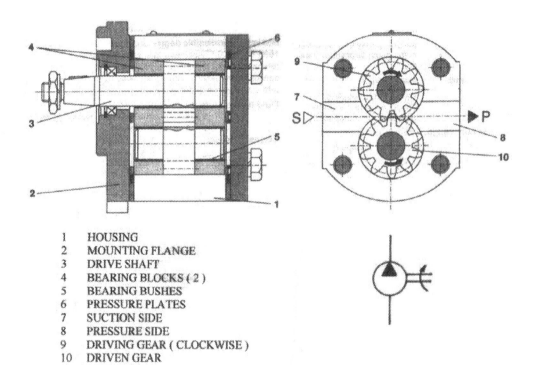

1	HOUSING
2	MOUNTING FLANGE
3	DRIVE SHAFT
4	BEARING BLOCKS (2)
5	BEARING BUSHES
6	PRESSURE PLATES
7	SUCTION SIDE
8	PRESSURE SIDE
9	DRIVING GEAR (CLOCKWISE)
10	DRIVEN GEAR

Illustration 25. External gear pump (Bosch Rexroth)

The gear pump is simple in concept and is manufactured in a wide range of sizes and displacements. Fluid is carried from the in-port to the out-port between the teeth of both gears. The displacement is proportional to the gear width, tooth size and gear diameter. For instance pumps with a gear diameter of 35 mm may have gear widths of 12, 18, 24, 30 and 36 mm, giving 6 different output flow rates.

Large output pumps will be physically larger in gear diameter and width.

There is a trade off between the number of teeth and the pump output. The more teeth there are the smoother the flow but, of course, to obtain high flows larger diameter gears need to be used. Experience indicates that the optimum number of teeth is about 12.

Trapping of fluid between the gears as they pass the mesh point can generate high, localised pressures which when released creates noise. Channels in the side plates are precisely machined to alleviate these pressures.

The volumetric efficiency of gear pumps i.e. the proportion of actual output flow relative to theoretical flow is affected by :-

a) leakage between the gears as they pass the mesh point
b) leakage past the side of the gears back to the inlet port
c) leakage over the top of the gears i.e. between gear outside diameter and pump body back to the inlet port.

Over the years improvements in the quality of machining of gears and bodies and the use of pressurized side plates which counteract the forces on the body caused by the outlet pressure all contribute to efficiency improvements to the point that gear pumps are both efficient and cost effective.

Gear pumps are used in many applications from very low power i.e. boat autopilots to high power, construction, mobile and agricultural machinery.

The output pulses due to the way the teeth mesh and this can result in noisy operation. Pumps are available with helical gears to provide smoother flow and reduce noise levels.

FILTRATION ISO4406 19/17/13
VISCOSITY 20-60cSt optimum

WHAT CAN GO WRONG

Fluid must be clean, good quality and correct viscosity for long pump life (see Chapter 4.)

Damage can occur by scoring the inside of the pump body where it seals against the gears causing loss of output flow.

Cavitation i.e.voids in the fluid caused by vacuum created at the pump inlet by restrictions in the inlet piping can cause noise and erosion of pump parts. The voids can collapse violently as they are subject to pressure.

Air bubbles entrained in the fluid have the same effect.

The bearings are subject to side loads due to the imbalance of pressure on the gears so bearing life is a factor especially if the input shaft is loaded by the drive arrangements.

Excessive side loads in the input shaft due to misalignment of the drive can reduce bearing life significantly.

Shaft seals can fail although this is unlikely as there is little pressure on them when the pump is used in open circuit. The external lipseal on the shaft stops air being sucked into the inlet port.

Very small capacity gear pumps may have low pressure capability as the leakage losses can equal the pump output. This is a factor when they are used in power packs when driven by DC motors which slow down as the load increases.

INTERNAL GEAR PUMPS

The pumping element is an internal and external gear meshing together the drive gear being the internal one. Fluid is carried from the inlet port to the outlet port between the gear teeth. There are two common types:-

1. The gears are separated by an insert as per illustration 26.
2. The internal gear has one less tooth than the external gear as per illustration 27.

As with external gear pumps the efficiency is dependent on the quality of the gears and the clearance the major advantage of this type of arrangement is a smooth output and quiet operation.

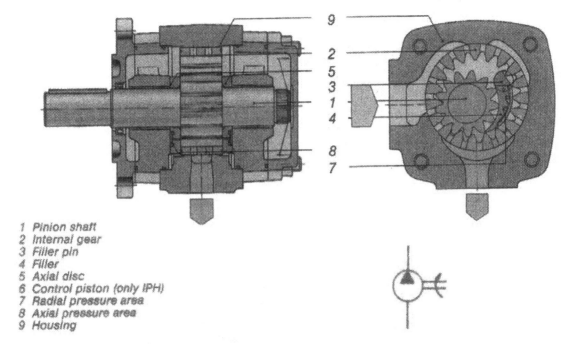

1 Pinion shaft
2 Internal gear
3 Filler pin
4 Filler
5 Axial disc
6 Control piston (only IPH)
7 Radial pressure area
8 Axial pressure area
9 Housing

Illustration 26. Internal gear pump (Voith)

Illustration 27. Internal gear pump (Gerotor type)

WHAT CAN GO WRONG

Fluid must be clean, good quality and correct viscosity for long pump life (see Chapter 4.)

Damage can occur by scoring of the body where it seals against the gears causing loss of flow. Cavitation or aeration of the fluid due to poor inlet conditions can cause damage resulting in loss of flow.

The bearings are subject to loading due to the imbalance of pressure on the gears so bearing life can be a significant factor.

Excessive side loads on the drive shaft due to misalignment of the drive can reduce bearing life. Shaft seals can fail although this is unlikely as there is little pressure on them.

FILTRATION ISO4406 19/17/13
VISCOSITY 20 –60cSt optimum

VANE TYPE PUMPS

The Vickers Vane Pump developed in the early part of the 20th century provided the power to create a revolution in the machine tool industry and the catalyst for the widespread use of hydraulic power for machine tool and other industrial machinery.

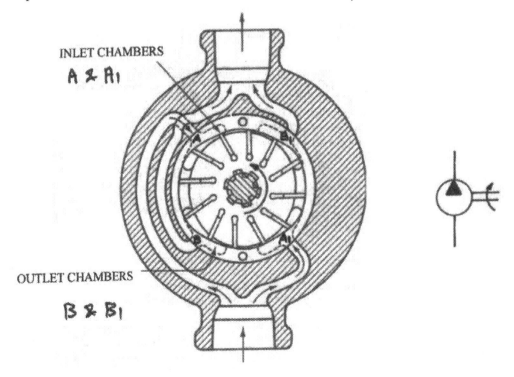

Illustration 28. Vane pump operating principles (Eaton)

This concept has lead to the development of vane type pumps for many other applications including construction machinery.

The vane pump displacement is dependent on the "throw" of the internal cam form and the width of the cam.

The force on the vane tip exerted by centrifugal force and pressure is controlled by a combination of springs and vane design, the output pressure being applied to only a small area of the vane sufficient to keep it in contact with the cam form.

The cam forms are a compromise to provide a smooth output flow and at the same time keep the acceleration forces on vanes in and out of their slots to a minimum.

Leakage past the rotor faces etc. is collected in the body and returned to the inlet port so no external drain is required.

The volumetric efficiency of the vane pump is affected by leakage past the vanes and leakage past the sides of the rotor.

Highly accurate machining of the component combined with floating side plates which combat the tendency of the pressure to push the sides apart have resulted in significant improvements. Vane pumps are also constructed to provide variable flow. This is achieved by moving the cam ring relative to the rotor.

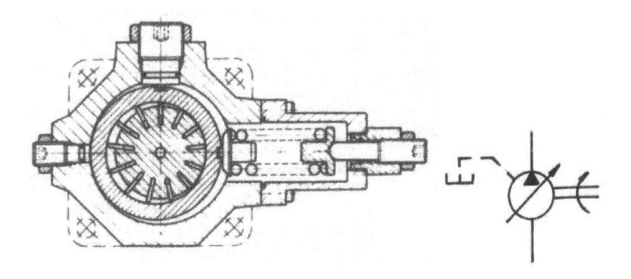

Illustration 29. Variable delivery vane pump (Eaton)

Vane pumps are used successfully in industries including machine tool, municipal, mobile, construction, injection moulding and die casting.

WHAT CAN GO WRONG

Fluid must be clean, good quality and correct viscosity for long pump life (see Chapter 4.)

Damage (ridging) can occur to vanes and cam rings due to restriction of the inlet causing cavitation i.e. voids in the fluid which collapse when exposed to pressure causing shock.

Scoring of the vane tips and cam rings and also the side plates due to debris in the fluid will result in loss of flow.

Aerated fluid will have the same affect as cavitation.

Due to the balanced design of the fixed delivery pump no significant loads are imposed on the bearings by the pumping action. There are side loads on the variable delivery pump bearings due to the pressure imbalance across the rotor.

Misalignment of the drive shaft with the pump input shaft can cause high side loads with significant reduction in bearing life. Flexible coupling can impose high side loads if the shafts are not aligned correctly.

Shaft seals can fail although this is unlikely as there is little pressure on them. The external lip seal on the shaft stops air being sucked into the inlet port

FILTRATION ISO 4406 18/16/13
VISCOSITY 20 – 60cSt optimum

PISTON PUMPS

The act of causing a piston to reciprocate in a bore in combination with inlet and outlet non-return valves forms the basis of the original piston type pumps. They now come in several forms some of which use non return valves, others do not.

RADIAL PISTON PUMP

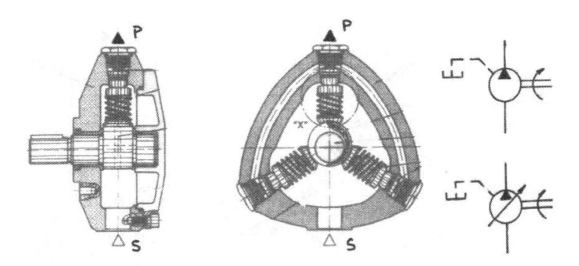

Illustration 30. Radial piston pump (Bosch Rexroth)

The displacement is a function of piston area and piston displacement. Relatively expensive compared to gear and vane pumps they provide long life with quiet operation and are mainly used for high pressure applications in industrial machinery.

A similar type of pump in which the pistons are located in a central piston block with an external cam ring can be used to provide variable displacement by moving the cam ring relative to the shaft.

FILTRATION ISO4406 18/16/13
FLUID VISCOSITY 20 – 60cSt optimum

AXIAL PISTON PUMP

The pumping elements (pistons) are in line with the drive shaft.

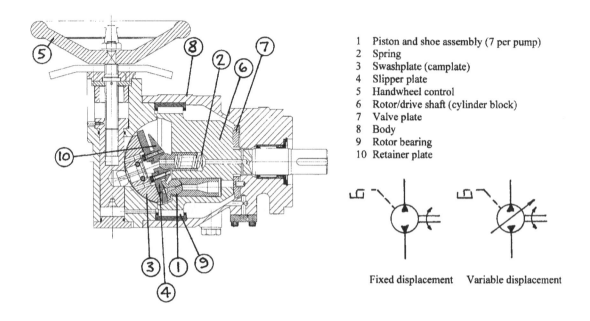

1   Piston and shoe assembly (7 per pump)
2   Spring
3   Swashplate (camplate)
4   Slipper plate
5   Handwheel control
6   Rotor/drive shaft (cylinder block)
7   Valve plate
8   Body
9   Rotor bearing
10 Retainer plate

Fixed displacement   Variable displacement

Illustration 31. Axial piston pump (Rotary Power)

As the rotor (piston block) rotates the pistons are constrained to move in and out by the angle of the swash (cam) plate and the "retainer" plate.

The piston shoes are swaged or crimped onto the ball end of the piston to provide enough clearance to be able to rotate but with minimum of axial "play".

The rotor is kept in contact with the valve plate initially by the spring, then by system pressure.

When the pistons are on the out stroke they draw fluid into the space created through the inlet port. On the in stroke the fluid is forced into the outlet port.

The piston shoes have a balancing area in contact with the slipper plate which is connected through the piston to the port. This counterbalances a large proportion of the force of the piston on the plate created by the output pressure on the output stroke.

Axial piston pumps have odd numbers of pistons either 7 or 9 to assist in reducing the output ripple or variation in flow as each piston delivers fluid into the outlet port.

Output flow from the pump is pulsating because the combined output of the pistons, as they deliver fluid into the outlet port, is not constant. The changeover from pressure to suction can generate noise. This is controlled by the use of tapered slots at the edge of the ports.

The volumetric efficiency of the pump depends mainly on the piston clearances and leakage across the port face and piston shoes. Generally this leakage is small and the axial piston pump is reputed to have excellent volumetric efficiency.

By rotating the swash (cam) plate the pump can be used to provide variable delivery and in conjunction with a piston motor can be used in closed loop transmissions i.e. the return from the motor is directed to the inlet of the pump and by moving the plate across centre the motor can be reversed.

The piston pump is exceptionally versatile and can be provided with a wide range of control options some are:-

     Pressure Compensated
     Constant Power
     Manual
     Servo
     Etc.

Axial piston pumps have limited ability to self aspirate i.e. draw fluid from a tank and when required to do so are limited in drive speed. The shock loads caused by cavitation or aeration of the fluid shorten the pump life.

When used with a boost pump or in a closed loop system the drive speed can be much higher. The success of the axial piston pumps in mobile, mining, construction, agricultural and marine industries is due to the ability to be driven at the same speed as the prime mover i.e. internal combustion engine when part of a closed loop system.

Leakage past the pistons port face etc. is collected in the pump casing and this is piped back to the reservoir in such a way that the pump casing remains full at all times.

For cooling purposes in closed loop systems the excess flow of fluid from the boost supply can be directed through the pump body and then to the reservoir.

Constant pressure systems using axial piston pumps with pressure compensation i.e. the pump only provides a flow on demand is used successfully on agricultural tractors amongst other applications.

Small capacity axial piston pumps with over centre i.e. reverse flow capability are used successfully on golf course grass cutting machines to provide variable vehicle driven motion both forward and reverse without a gear box.

FILTRATION ISO 4406 18/16/13
FLUID VISCOSITY 20 – 60cSt optimum

WHAT CAN GO WRONG

Fluid must be clean, good quality and correct viscosity for long pump life (see Chapter 4.)

Bearings maintain the cylinder block and shaft in position and there is considerable load on the bearings imposed by the pumping action. This is taken into account in the pump design but may eventually limit the life of the pump.

Flexible couplings can impose high side loads on the drive shaft if the shafts are not aligned correctly which can add to the bearing loads.

Shaft seals can fail if subject to excessive pressure but are unlikely to do so.

Scoring of the port faces and shoe faces due to debris in the fluid will cause high internal leakage. Wear between pistons and bores due to the side loads imposed on the pistons will result in internal leakage and a drop in efficiency, wear between pistons and shoes will result in reduced stroke and high leakage.

High standards of filtration are essential.

BENT AXIS PISTON PUMPS

Similar in principle to the axial piston pump, in the bent axis pump the piston block is offset at an angle to provide the piston displacement i.e. pumping action.

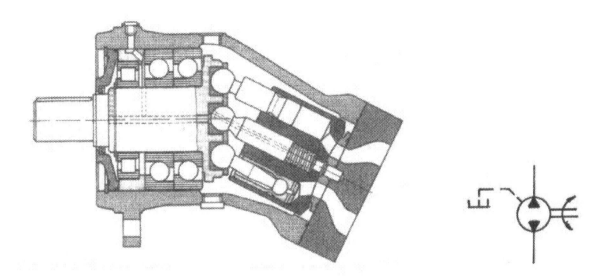

Illustration 32. Bent axis piston pump (Bosch Rexroth)

This type of pump has similar function to an axial piston pump. It is used in the same sort of applications.

FILTRATION ISO 4406 18/16/13
FLUID VISCOSITY 20 – 60 cSt optimum

WHAT CAN GO WRONG

The same things can go wrong as with axial piston pumps, see previous section.

11

MOTORS

Hydraulic motors provide a wide range of speeds and torques but none have the capacity to hold a load without the use of a brake. Commonly used types are described in this section.

GEAR MOTORS

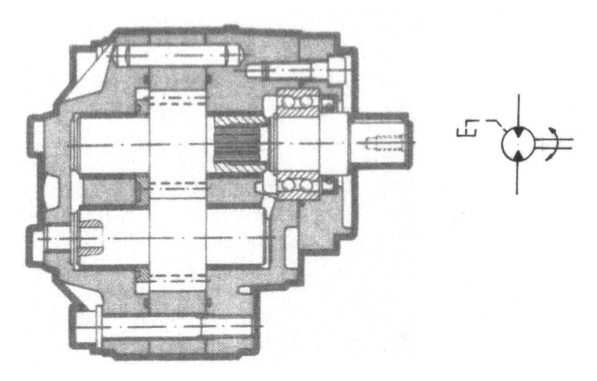

Illustration 33. External gear motor (Eaton)

External gear motors work in the opposite way to gear pumps. Fluid is introduced at a service port. Pressure is applied across the area of the gear teeth of both gears which causes the gears to rotate. One of the gears forms part of the output drive shaft.

The output torque is proportional to the pressure applied to the gear sizes and width and the speed is proportional to the flow introduced.

The operation of external gear motors at low speed tends to be uneven and most manufacturers recommend a minimum speed of 500 rpm.

The breakout or starting torque under load is in the order of 65% of the running torque.

To avoid pressurizing the shaft seal a drain port is normally provided which allows leakage past the side faces to be returned to the reservoir.

Gear motors can rotate in either direction depending on which port is subject to pressure fluid and are used in many types of industrial municipal, agricultural and construction machines.

FILTRATION ISO 4406 19/17/13
FLUID VISCOSITY 20 –60cSt optimum

WHAT CAN GO WRONG

Fluid must be clean, good quality and correct viscosity for long motor life (see Chapter 4.)

Damage can occur by scoring the inside of the motor body where it seals against the gears or where the gears seal against the side faces.

The bearings are subject to loading due to the imbalance of the pressure on the gears so bearing life is a factor especially if the shaft is side or end loaded by the drive arrangement.

Shaft seals can fail due to pressure build up in the motor body if the drain line is restricted.

Overrunning loads can cause cavitation and loss of control.

GEAR MOTORS AS FLOW DIVIDERS

To operate two or more systems at the same speed from one pump the shafts of gear motors can be coupled together in any number. The supply is connected to all the inlet ports and each outlet port is connected to a separate system.

There is a small pressure drop across each motor but this is an efficient and accurate way of dividing flow.

This type of flow divider can also be used as a pressure intensifier.

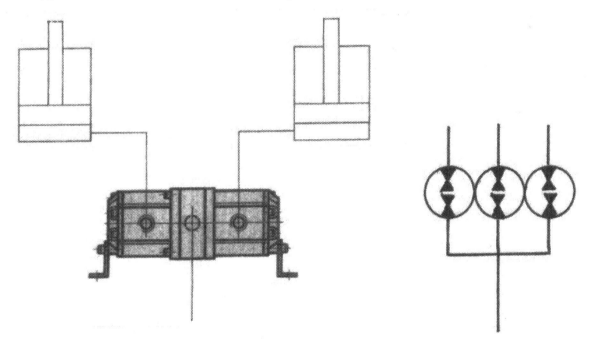

Illustration 34. Gear motor flow divider (Casappa). The symbol is for a flow divider with 3 flow paths.

WHAT CAN GO WRONG

Fluid must be clean, good quality and correct viscosity for long motor life (see Chapter 4.)

Damage can occur by scoring the inside of the motor body where it seals against the gears or the gears seal against the side faces.

Shaft seals can fail due to pressure build up in the motor body if the external drain line is restricted.

61

VANE MOTORS

These are of similar construction to vane pumps.   Pressurized fluid is provided to the vanes which causes the rotor to rotate in either direction The vanes are initially spring loaded into contact with the cam ring. When operating the vanes are kept in contact by a small pressure imbalance between the top and bottom of the vane.

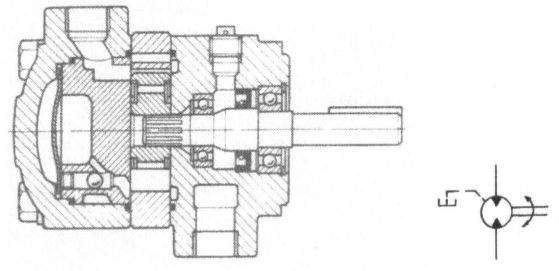

Illustration 35. Vane motor (Eaton)

The output torque is a function of cam displacement and width. To avoid pressurizing the shaft seal an external case drain is provided.

The breakout torque or starting torque under load is in the order of 70% of the running torque.

The operation at low speed tends to be uneven and most manufacturers recommend a minimum speed of 100 rpm.

WHAT CAN GO WRONG

Fluid must be clean, good quality and correct viscosity for long motor life (see Chapter 4.)
Overrunning loads can cause cavitation with loss of control.
Scoring of the cam ring vane tips and side faces will result in loss of efficiency (high leakage).
High side loads on the shaft will reduce bearing life.
Restriction in the drain life could cause shaft seal failure.

AXIAL PISTON MOTORS

Of similar construction to axial piston pumps they are generally of fixed capacity as the efficiency tends to fall off as the swash angle is decreased. Axial piston motors have an odd number of pistons to reduce the ripple effect on the output speed.

As in axial piston pumps the cylinder block is pressed against the port face by a combination of the spring and the effect of pressure on the pistons which is partially balanced by the grooves in the port face.

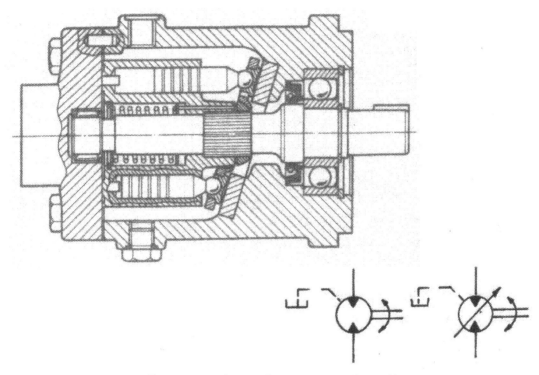

Illustration 36. Axial piston motor (Eaton))

Axial piston motors are used in open loop or closed loop circuits in conjunction with axial piston pumps.

Leakage past the port faces and pistons is piped back to the reservoir in such a way that the motor casing remains full at all times. In closed loop systems a flow from the boost supply is passed through the motor to assist cooling.

The breakout torque or starting torque under load is in the order of 80% of the running torque.

The operation at low speed tends to be uneven and most manufacturers recommend a minimum speed between 50 and 100 rpm.

WHAT CAN GO WRONG

The problems are the same as those for axial piston pumps – please refer to the relevant section.

High standards of filtration are essential. Overrunning loads can cause cavitation with loss of control and reduced life.

BENT AXIS PISTON MOTORS

Of the same construction as bent axis piston pumps these motors are suitable for open loop or closed loop transmissions.

Illustration 37. Bent axis piston motor (Parker VOAC)

The same recommendations apply as for bent axis piston pumps.

INTERNAL GEAR (GEROTOR) MOTORS

This type of motor does not have its equivalent as a pump. It has a unique feature which combines to form both a rotary motion and reduction gear in one, providing output speeds from lower than 10 rpm to as high as 500.

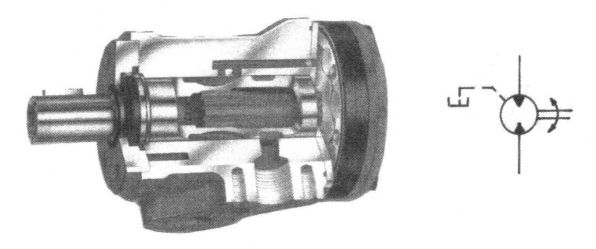

Illustration 38. Orbital motor (Sauer-Danfoss)

Pressure fluid is introduced at the inlet port which is connected through a valve to one side of the interface between the internal and external gear. The internal gear has one less tooth than the external gear, the outside of the internal gear being a close fit in the external gear. The result of the imbalance produces a torque and the transmission shaft which is connected to the internal gear rotates in the body in a planetary manner and drives the output shaft through a constant velocity joint

To avoid excessive leakage the gear parts are machined to very close tolerances. The manufacturing precision of the parts can affect the smoothness or otherwise of the rotary motion particularly at low speed.

The breakout or starting torque is a high percentage of the running torque.

The smooth operation at low speed is a feature of this type of motor and speeds down to less than 5 rpm can be achieved by very precise machining of the components.

The output torque is proportional to the width of the rotating elements and the diameter of them. Each frame size therefore is dependent on the component sizes with the range of torque in each size being determined by the width of the rotating parts.
.

Most designs have options of high pressure mechanical shaft seals which can be pressurized up to system pressure and incorporate check valves in the motor body which connect the drain port to the low pressure or outlet port to avoid having an external drain line. This feature allows motors to be used in series without loss of flow giving equal speeds from all motors in the series.

FILTRATION ISO 4406 19/17/13
FLUID VISCOSITY 20 – 60 optimum

These low to medium speed motors are used very successfully in all industries where rotary motion is required from yacht winches to agricultural machines, municipal and construction equipment. They are often combined with gearboxes to provide very low speed high torque output.

WHAT CAN GO WRONG

Fluid must be clean, good quality and correct viscosity for long motor life (see Chapter 4.)

Low pressure shaft seals can fail due to pressure build up if the drain line is restricted.

Scoring or wear can take place between the rotating elements and between the rotating elements and the body. The bearings are subject to side loads due to pressure imbalance on the rotating parts although they are not subject to high speed. Shaft bearing life may be a factor where the drive shaft has high side loads.

Overrunning loads can cause cavitation with loss of control and reduced life.

LOW SPEED HIGH TORQUE MOTORS

The workhorses of the construction, marine and heavy industries, these motors provide rotary motion from wheel drives to crane hoists amongst many others.

The method of generating torque is generally by pistons equally spaced around either an internal or external cam form. The pistons are pressurised in turn to provide a turning moment. Very high torques can be achieved.

The breakout torque is a high percentage of the running torque.

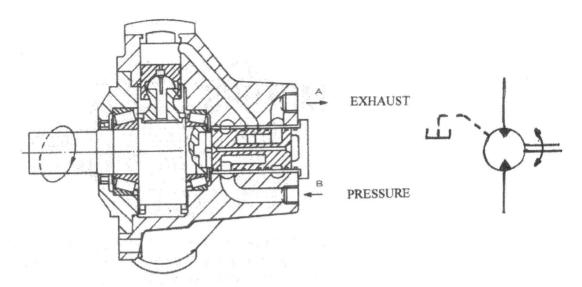

Illustration 39. High torque low speed motor (Kawasaki) (Clockwise rotation shown, looking on shaft end.)

WHAT CAN GO WRONG

Fluid must be clean, good quality and correct viscosity for long motor life (see Chapter 4.)

Scoring of piston/bores and port faces will reduce efficiency by causing excess leakage.

Scoring of cam forms can occur between piston and cam.

Because of their nature high side loads can be applied to the motor output shaft bearings which could affect their life.

Shaft seals can fail due to pressure build up in the motor body if the drain line is restricted.

FILTRATION ISO 4406 18/16/13
VISCOSITY 20 – 60 cSt optimum

12

CYLINDERS

The traditional way of converting hydraulic power back to mechanical power, cylinders encompass a wide size range. They are either double or single acting and all function in the same way.

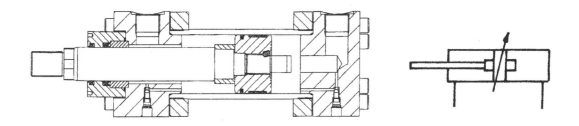

Illustration 40. Double acting cylinder with cushioning at both ends

Pressure applied to the full bore area of the cylinder produces a force which is the area of the cylinder times the pressure applied.

Force = Pressure x Full Bore Area   (bar x cm²)

The use of the bar for pressure does not lend itself readily to appreciate the forces involved.

However multiply by 10 to obtain the force in Newtons – (1)
Divide (1) by 4.45 to obtain the force in lbs.
Divide (1) by 9.8 to obtain the force in kgs.

When pressure is applied to the rod end the force produced is the full bore area minus the rod area times the pressure applied.

Force = Pressure x (Full Bore Area – Rod Area)

If the same pressure is applied to both ends at the same time the effective area is the area of the rod and the extending force produced the rod area times the pressure applied (see regenerative system).

Force = Pressure x Rod Area

In some applications to equalise the forces and speed in each direction the cylinder is fitted with a shaft at both ends e.g. yacht steering etc.

REGENERATIVE SYSTEM

Another way of obtaining equal forces and speeds is to arrange for the cylinder rod area to be 50% of the full bore area. When pressure is applied to the rod end a force and speed is obtained. The same force and speed is obtained in the other direction by connecting both the rod end and the full bore end together. The flow from the rod end supplements the flow into the full bore end.

This is known as a Regenerative System. It can also be used with any double acting cylinder, to obtain for instance a fast approach then switch to conventional operation for full output force when in position i.e. baling press etc.

In the regenerative mode the cylinder extending speed is determined by dividing the fluid flow rate by the rod area. The force available is the pressure times the rod area.

CUSHIONING

To avoid high shock loads when a cylinder piston reaches the end of its stroke a cushion can be incorporated. This is formed by an extension of the piston entering a matching bore in the cylinder body. The fluid is trapped in this area and constrained to flow out through a restrictor slowing the piston down to an acceptable level. The restrictors can be in the form of adjustable orifices which enable the cushioning effect to be adjusted. Cushions slow the piston down but do not reduce the force.

BUCKLING

For a cylinder that is working in compression i.e. pressure being applied to the full bore end the rod transmitting the force to the machine the rod must have sufficient strength to avoid the effects of buckling or bending.i.e. the rod diameter must be as large as possible The material strength and the length of the rod are factors.

$$\text{Buckling Load} = \frac{\pi^2 \times E \times I}{\text{Buckling Length}^2}$$

$$\text{Where } I = \text{Moment of inertia of rod} = \frac{\pi d^4}{64}$$

This is divided by a factor of safety to determine the safe working load.

Refer to Principles of Hydraulic System Design published by Coxmoor Publishing Company for calculations, and BFPA Engineers Data Book for maximum rod lengths and strokes.

Cylinders working in tension do not have the tendency to buckle so it is desirable from this point of view to have long thin cylinders working in tension.

Various mounting methods are shown below:-

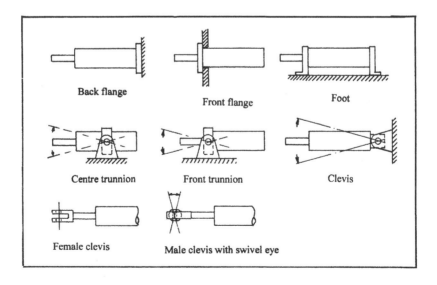

Illustration 41. Methods of mounting cylinders

METHODS OF CONSTRUCTION

The base tube can be fixed to the cylinder ends by bolts, screwed in, welding and studs. Each method has its advantages and disadvantages. Cylinders which are welded together can only be serviced by replacement cylinder. Other types can be disassembled and can have seals bearings etc. replaced.

Cylinders can now have positional sensors built in to provide feedback for electronic control system of both position and velocity.

MATERIALS

Steel is the predominant material used. The cylinder bore is finished by honing to a very high standard of roundness and surface finish. (See Seals chapter).

The rods of cylinders for outdoor use are chrome plated before finish grinding to obtain a finish which will resist corrosion. The degree of sophistication of this rod finish is determined by the application i.e. cylinders used in a marine environment must be corrosion proof for many years.

Where cylinders are exposed to impact damage i.e. automotive tailgate lifts etc. gaiters may be fitted to protect the rods.
Any damage to rods due to corrosion or impact will cause damage to the seals as the rods slide in and out.

CYLINDER SEALS

Seal manufacturers have developed sophisticated seals for cylinders mainly in nitrile rubber or polyurethane. Many cylinders have a requirement to be held in position without movement i.e. no leakage past the seals as in cranes etc. The ability of cylinders to do this is almost taken for granted but the troubleshooter must be aware that seal damage can occur with subsequent leakage.

SEAL FRICTION

Seals by their nature must be compressed to be effective. This results in force being needed to overcome the initial seal friction when starting to move and also to keep the piston moving. Whilst this force is small compared to the force generated by the cylinder allowance (in the form of extra pressure) should be made for this when calculating cylinder sizes at the design stage.

Rapid cylinder speed needs seals which will tolerate high speed.

SAFETY

Where it is vital that a cylinder does not move over a period of time mechanical means of holding the machine in place is essential.

WHAT CAN GO WRONG

The most obvious cause of failure of the rod seal is damage to the exposed part of the rod which then retracts past the seal and tears it.

Rod damage can be caused by impact of stones or other debris and by corrosion. Many cylinder rods are exposed to salty wet conditions which unless taken into account at the outset will cause corrosion and subsequent damage to the rod seal.

The piston seals can be scored due to debris and since they are at the ends of the pipe any debris due to careless pipe fitting or other factors will inevitably finish up in the cylinder.

For detailed information on the use of seals in cylinders please refer to Chapter 15.

Side loads on the rod can cause wear to the rod and bearing. See previous section about buckling and bending of cylinders in compression.

Where an "eye" is fitted to the end of a piston rod the point where the rod is reduced in diameter to provide a threaded extension can be at a weak point especially if subject to side loads.

13

ROTARY ACTUATORS

Used for rotary movement of less then 360°, they can be vane, rotary cylinders or a combination of piston and multiple start thread (helical spline) as per illustration.

Illustration 42.   Rotary actuator (Sauer-Danfoss)

FILTRATION ISO 4406 20/18/15
VISCOSITY 15 – 200 cSt

14

VALVES

The general filtration requirements for valves are:

Directional, pressure, flow controls ISO4406 18/16/13
Proportional valves   ISO4406 17/15/12
Servo valves   ISO4406 16/14/11

VISCOSITY   13 – 500 cSt

Control of pressure is a fundamental requirement of all hydraulic systems since lack of control can result in overstressing and breakage of system components.

RELIEF VALVES

The basic relief is a ball which is pressed into a hole by a spring.

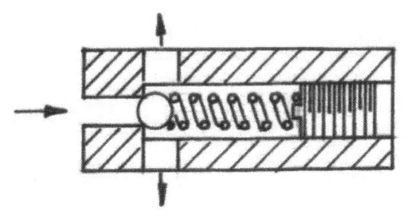

Illustration 43. Basic relief

The ball will lift and allow flow past it when the pressure times area equals the spring force.

This arrangement has limitations. The ball can become unstable. The fluid velocity past it and the spring rate can result in unacceptable characteristics.

Ideally flow past a relief valve should start at a specific pressure (cracking pressure) and that pressure should be maintained regardless of the flow rate. There should not be any instability (noise) and the valve must respond to a rise in pressure instantaneously. The valve should close (reseat) at a pressure very close to the cracking pressure. To obtain these characteristics something more sophisticated than a ball and spring is required.

For lower flow rates this can be direct acting and for higher rates two stage i.e. pilot or control stage and high capacity main stage.

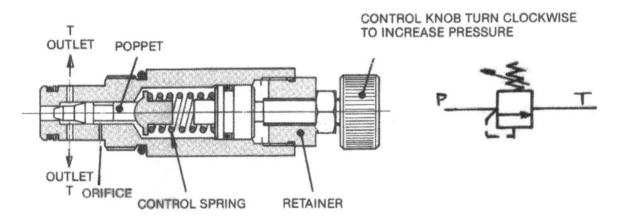

Illustration 44 Direct acting relief (cartridge)

This valve is a direct acting poppet type adjustable pressure relief valve controlling the pressure in the inlet port. The outlet port is used as the exhaust. System pressure acts on the end of the poppet. When the force created by the pressure acting over the poppet seat area exceeds the spring force (valve setting) the poppet lifts allowing flow from the inlet port to the outlet port. The area of the annular shoulder on the spool and the restriction of the outlet ports creates extra poppet lift as the flow increases to produce the flat characteristic curve. (See illustration 45). The orifice connecting the spring chamber to T provides damping.

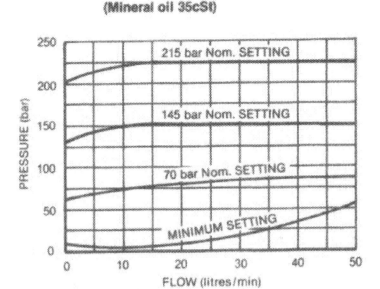

Illustration 45 Characteristic curves direct acting relief (Ill.44)

When more sophistication is introduced relief valves make the transition to pressure controls.

PRESSURE CONTROLS

Pressure controls operate to control system pressure continuously for instance in a constant pressure machine tool system.

For this purpose the valves need to have rapid response, smooth control, a flat characteristic i.e. little change of pressure with flow rate. Normally two stage valves with hardened steel seats and poppets are used for pressure controls to obtain the long life required.

Some leakage between the P&T ports is present in most two stage designs (i.e. around the main poppet outside diameter). Although small this is undesirable if a load holding function is required.

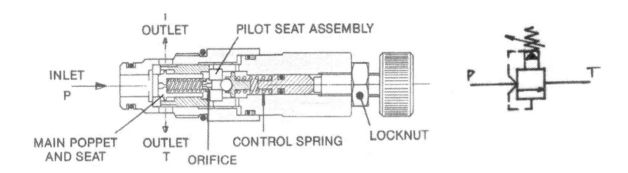

Illustration 46. 2 stage pressure control (cartridge)

The valve controls the pressure in the inlet port. The outlet port is used as the exhaust port. At pressure up to the valve setting, the pressure in the inlet port and the pressure in the chamber above the main poppet are the same. As pressure rises past the valve setting the pilot relief lifts and allows fluid flow to the outlet (exhaust). The fluid flow causes a pressure difference across the orifice in the main poppet. When the fluid flow through the orifice rises to a point where the pressure difference across it is sufficient to move the poppet against the poppet return spring, the poppet opens and allows flow of fluid from the inlet port to the outlet port via the main seat. When the sytem pressure falls the pilot valve closes reducing the flow through the control orifice to zero and the main poppet closes.

Under the steady state conditions when the valve is being used as a pressure control the pilot assembly and main poppet take up intermediate positions allowing continuous flow at constant pressure from inlet to outlet ports.

The flange on the main poppet provides a flow resistance which provides "lift" and increases the poppet opening, this effect produces a flat characteristic curve as shown in illustration 47.

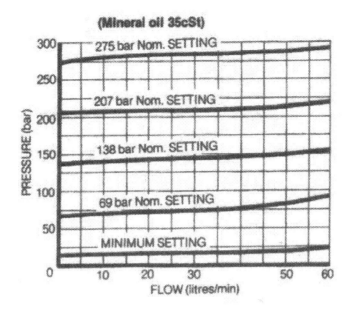

Illustration 47. Characteristic curves 2 stage pressure control (Ill. 46)

CROSS LINE RELIEF VALVES

Relief valves are often used to control the maximum pressure in actuators both for normal motion and for braking. To avoid cavitation the exhaust flow from the valve is piped into the other connection to the actuator. The valve has to be capable of accepting system pressure on the exhaust port or a check valve is fitted in series to avoid malfunction.

Where double acting cylinders are used provision has to be made to cope with the different flow rates due to the difference in flows generated by the different areas of the cylinders.

Pressure relief valves are also used to control rise in pressure due to expansion of fluid as temperature rises – see Temperature Chapter 19.

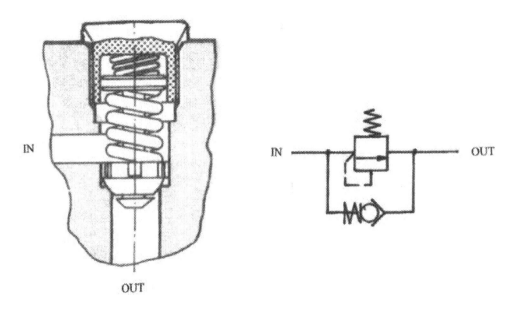

Illustration 48. Preset relief and check valve (Parker VOAC)

The valve shown in illustration 48 is a preset relief, the centre poppet provides the relief function. This allows flow from IN to OUT when the pressure reaches the setting of the main (large diameter) spring. The whole valve moves back when flow is from OUT to IN against the smaller spring in the cap which offers little resistance. This simple but effective device has many applications in mobile systems.

Relief valves provide essential pressure protection of system components and are used widely in systems to control the maximum pressure from the pump and also in other parts of the circuits where there is a requirement to control the pressure locally i.e. when a motor is stopped by a directional valve with blocked service ports. A relief valve could be used to limit the pressure caused by the motor/load inertia i.e. protect the motor, pipes, valves etc. from excess pressure.

Any part of a system which can be subject to an induced uncontrolled pressure in excess of the design pressure of the components should be protected by a separate relief valve.

WHAT CAN GO WRONG

Springs in direct acting valves can be on the limit of stress and they can 'set' over a period of time reducing the valve pressure setting.

The constant flow between seat and poppet of pressure controls can cause erosion and consequent high leakage between P & T ports. The higher the pressure setting the more likely this is to happen.

Debris in the fluid can get stuck between poppets and seats causing high leakage between P & T ports.

Single stage relief valves often have steel seats which are not hardened. When used as a pressure control i.e. continuous operation, the life can be quite short due to erosion. The pressure setting in this case has a significant effect on the life, the lower the setting the longer the life.

PRESSURE REDUCING VALVES

Used where a lower pressure is required in part of a circuit other than the main circuit, the pressure reducer can be single stage which are normally used in low flow / low pressure applications. For higher flows / pressures two stage valves

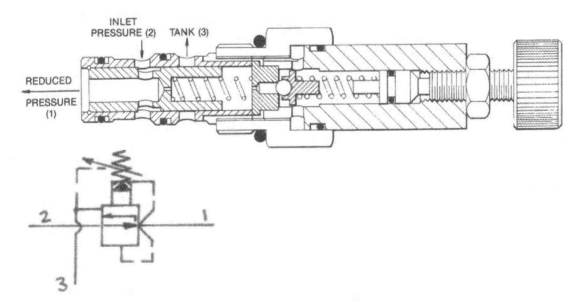

Illustration 49. Pressure reducing valve (cartridge)

The valve controls pressure in the reduced pressure port (1), the inlet connection being to the pressure port (2) and the tank port (3) being normally connected to exhaust at atmospheric pressure.

The spool is normally open allowing flow from the pressure port to the reduced pressure port. At reduced pressures below the setting of the pilot valve the pressure in the chambers above and below the orifice in the spool are the same. As the pressure in the reduced pressure port rises to the setting of the pilot valve the pilot valve lifts and allows flow to tank. The pilot flow causes a pressure difference across the orifice in the spool sufficient to move the spool against the return spring and shut off the flow from the pressure port to the reduced pressure port.

All of the time that the reduced pressure is at the pilot valve setting the spool opens enough to allow sufficient flow to keep the reduced pressure constant. When the reduced pressure falls the pilot valve closes and the spool opens fully.

If the reduced pressure rises above the setting of the pilot valve the spool travels further and allows the reduced pressure port to relieve to tank.

When functioning as a reducer the power loss due to the flow at the pressure drop across the valve is converted directly into heat.

WHAT CAN GO WRONG

Pressure reducer components are designed and manufactured for long life. Debris can cause the main operating element either poppet or spool to stick.

Each valve has its flow limits which if exceeded may cause malfunction.

PROPORTIONAL PRESSURE CONTROLS

The pilot stage of pressure controls can be replaced by a valve incorporating a proportional solenoid. The output force of the solenoid on the pilot cone or poppet is proportional to the current supplied to the coil.

ISOLATOR VALVES

When parts of a circuit need to be shut off, for instance to enable manual operation in the event of a power failure or to remove components for service, isolation can be achieved by using shut off valves i.e. cone on seat or ball valves. Ball valves are particularly easy to use as the handle only needs a turn of 90° to go from fully open to fully closed but the seating element is normally plastic so they do not have any metering capability.

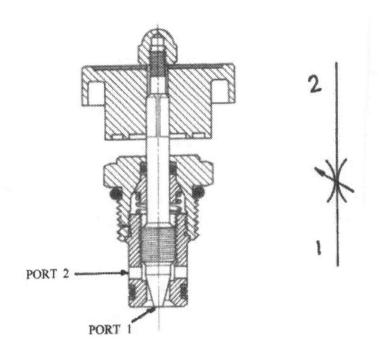

PORT 2 →

PORT 1 →

Illustration 50. Shut off valve (needle flow control) cartridge (Eaton)

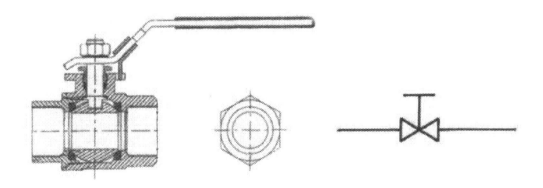

Illustration 51. Ball valve line mounted (Isis)

DIRECTIONAL VALVES

Directional valves are a basic requirement for most systems controlling the direction of movement of cylinders motors etc. There is such a wide range of sizes, operators, spools etc. that it is impossible to cover every aspect so a certain amount of generalisation has been applied.

SPOOL TYPE DIRECTIONAL VALVES

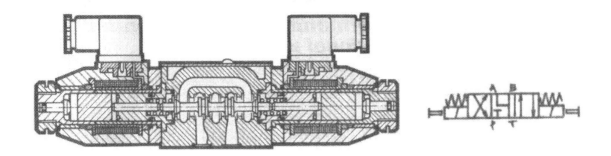

Illustration 52. Double solenoid valve (Eaton)

They consist of a machined or cast body normally cast iron with a precision bore finished to very close tolerances for diameter roundness and straightness and small axial tolerances for land positions. They are fitted with a hardened alloy steel spool again machined to a very high standard for land positions and close tolerances for diameter, roundness and straightness. The spool may also have shaped recesses to reduce the effect of the axial forces on the spool caused by fluid flow. Balancing grooves distribute pressure evenly around the spool.

HYDRAULIC LOCK

 Without balancing grooves on spools the pressure difference laterally across body lands can be distorted by slight variations in geometry causing an imbalance of radial forces on the spool which results in friction causing the spool to stick. This phenomenon is known as "hydraulic lock" and was much more prevalent when bores where honed using hand tools. Balancing grooves are used where there is sufficient axial spool overlap to enable them to be accommodated.

Illustration 53 describes this phenomenon.

The imbalance of forces due to the different pressure gradients results in the spool being forced into contact with the body.

Balancing grooves around the spool allow the intermediate pressures to be evenly distributed around it, this combined with good spool and body geometry eliminates this problem. The grooves need to be narrow as any reduction in land length increases   fluid leakage past the land. (See chapter 3).

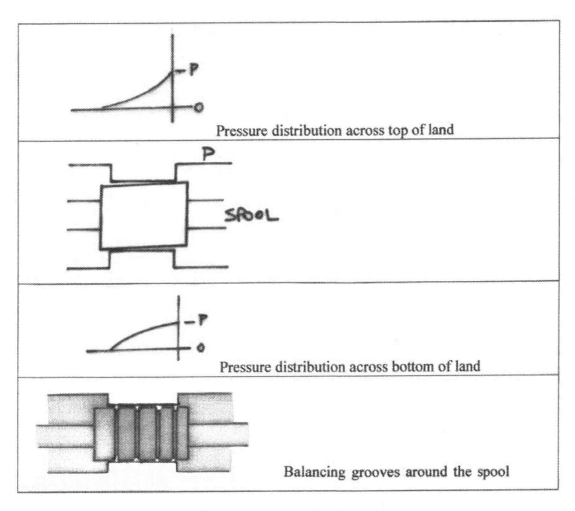

Illustration 53. Hydraulic Lock

LEAKAGE PAST SPOOLS

Because fluid leakage through an annular gap varies with the cube of the gap and leakage at best is wastage of energy and at worst allows loaded actuators to move the gap or clearance is kept to an absolute minimum. Manufacturers approach this in different ways. One may manufacture all bodies and spools to specific tolerances then select parts by measurement only if ultra low leakage is required. This allows spool interchangeability which is useful for distributors who may want to change a spool type.

Another may grade bodies and spools by diameter and select combinations which provide a specific range of clearance every time.

Another may measure the body bore and have facilities for finishing the spool diameter to give a specific very small clearance.

The cast iron used for bodies is of a very high quality in terms of strength and permeability and the introduction of zircon sand cores in the 1960s enabled castings with precision core positions and sizes to be produced economically.

Spools are machined from alloy steels which do not distort during hardening and provide a durable long life.

Virtually all solenoids in use today have the armature working in the hydraulic fluid enclosed by a tube which retains the armature and fluid.   The coil slips over the tube and is a close fit.

The armature is a close sliding fit in the thin walled tube which incorporates in its construction a non-magnetic section to concentrate the electromagnetic forces in the armature. Please refer to illustration 52.

This type of construction ensures trouble free operation and long life.

DIRECT ACTING SOLENOID OPERATED VALVES

Each type of spool has its own functional limits. The action of fluid flow past the lands of a spool as it is operating creates flow forces on the spool. These forces can be balanced by the flow forces on another land or may resist or assist the solenoid. These forces are a function of the pressure drop across the gap and the flow rate, some measure of compensation i.e. the spool shape can reduce these forces.

The result is that whilst a valve may be rated at a flow capacity of 50 litres per minute each type of spool will have functional limitations either higher or lower than this nominal figure.

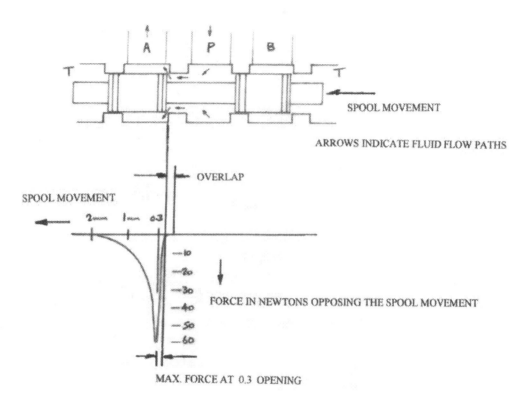

FLOW FORCES ON SPOOL AS PORTS "P" AND "A" ARE CONNECTED

Illustration 54. Flow forces on spool

This example is an 11mm diameter spool with a flow rate of 40 l/min the pressure at the 'P' port being 210bar. The pressure at the 'A' port is zero.

The point of maximum flow force occurs when the full flow of 40 l/min is flowing past the spool whilst the pressure drop across the spool is 210bar. As the spool opens further the pressure drop falls to a nominal figure and the flow force reduces. The forces on the spool are mainly caused by the increase in velocity of the fluid as it passes the spool land which causes a reduction in pressure on the spool land face. The pressure on the balancing land at the other side of the spool recess is higher so the effect is for this pressure imbalance to oppose the opening actuator.

Flow force is proportional to spool circumference and pressure but there are other factors that effect these forces such as the shape of the spool land and recess and the balancing effects of having another land opening at the same time i.e. B to T.

Ref. '*Improving the Performance of Directional Control Valves*' by D. Seddon Oct 1990.

Pressure drop across the valve will also vary depending on spool type, since the opening varies from spool type to spool type.

CONNECTORS

The industrial standard connector (ISO 4400) can be fitted with lights, diodes etc. to provide protection for the solenoid and indicate when it is energized. It does not, however, indicate if the valve has operated. Connectors may be different colours when marked with the letters 'A' or 'B' for identification.

Illustration 55. Solenoid connector (mPm)

Many other types of connector are used for economy or standardization with other parts of the control system.

DIRECTIONAL VALVE SYMBOLS

Two methods of showing the valve function are in use:

Method 1. Generally used in Europe the 'a' Actuator is always positioned at the same end of the valve as the 'A' port, 'b' actuator same end as 'B' port.

Method 2. Generally used in the USA the 'a' actuator is the solenoid which when energized connects port P to port A. The 'b' actuator when energized connects P to B.

METHODS OF IDENTIFYING ACTUATORS [SOLENOIDS]

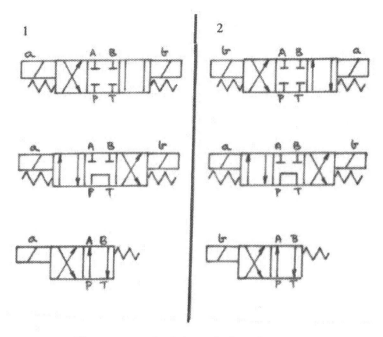

Illustration 56. Solenoid identification

What is common to the two systems is that the function obtained by energising the solenoid on the 'A' port end of the valve represents what happens physically, i.e. the spool movement relative to the body. The valves are essentially the same but in method 2 the a and b actuator can be either end of the valve and are marked on the valve nameplate accordingly.

The commonly used connecting electrical sockets for solenoids are marked A and B and are also different colours.

Anyone considering changing one valve for another should be aware that by keeping the solenoid references the same the function of the valve could be reversed.

Method 1 keeps the whole process of changing spools simple as long as anyone changing a valve or spool is aware that the valve function can change if a spool is changed.

Method 2 requires detailed knowledge of the spool function to determine which actuator is an 'a' and which is a 'b'. For some types of spool it could be either.

SOLENOIDS

Hydraulic valves use solenoids which are continuously rated, ie they can be energised continuously but AC directional valve coils are prone to overheating and burnout when subject to higher voltage than the coil is wound for. This used to be a bigger problem than it is these days as the standard voltages used in Europe were anything from 220 to 240V with a tolerance of ±10%. Basically AC coils needed to be wound for the specific voltage.

With the harmonisation of European voltages at 230V 50Hz (& 110v 60Hz in USA) the risk is reduced although if burnout of coils is occurring check the supply voltage. A DC coil with built in rectifier is more tolerant of AC voltage variations.

DC coils can present problems; for instance a 12 V DC coil may be expected to work at any voltage between 9 and 14.4 volts. A battery powered system with some voltage drop in the wiring may cause valve malfunction due to low voltage. On the other hand a 28 watt coil continuously energised at the alternator control voltage on an engine driven machine may be working at 14.4 volts giving a coil power consumption of 40 watts resulting in very high coil temperatures.

COIL CALCULATIONS

$W = I^2R$ & $I = \dfrac{W}{V}$ (Ohms Law)

Therefore $R = \dfrac{W}{I^2}$

W= POWER WATTS
I = CURRENT AMPS
R = RESISTANCE OHMS
V = VOLTAGE

For 28 watt 12 volt coil

$I = \dfrac{28}{12} = 2.333$ amps

Therefore resistance $R = \dfrac{28}{2.333^2}$

$= 5.14\ \Omega$

For 28 watt 24 volt coil

$$I = \frac{28}{24} = 1.1667 \text{ amps}$$

$$\text{Therefore resistance } R = \frac{28}{1.1667^2}$$

$$= 20.6 \ \Omega$$

Resistance will increase as the temperature increases and vice versa so the measurements are normally taken at 20°C.

The change of resistance of copper wire relative to temperature is 0.00393 per degree C i.e. approx 4% per 10°C. As the temperature rises the current and therefore the output force decreases at this rate.

To test that a coil is serviceable a multimeter can be used to test the resistance. This could be checked against the valve manufacturer's data. A typical figure would be 5.14 ohms for a 12 volt 28 watt coil at 20°C, and a 24 volt coil would have a resistance of 20.6Ω.

The inertia effect of current flowing through a DC coil can produce arcing across the control switch contacts when the coil is switched off reducing the life of the switch. This can be counteracted by using a free running diode in parallel with the coil. The response time of the valve may be extended by this when de-energising. This can be reduced by the use of a resistor about the same resistance as the coil as shown below.

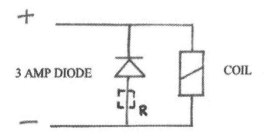

Illustration 57. Diode circuit for solenoid

WHAT CAN GO WRONG

Debris in the fluid can be trapped between the spool and body causing the spool to stick. Wear can take place between spool and body.

There is the possibility that the return spring could fail to move the spool if the manual override push pin in the solenoid tube becomes stuck in the 'in' position due to corrosion in outdoor applications.

"Hydraulic lock" and "silting" can cause spools to jam in the selected position especially when combined with fluid contamination. The higher the pressure and the longer the period of energisation the more likely this is to happen. Very fine filtration can help.

This type of failure can be dangerous when the valve in question is providing a safety function as the spool may fail to move when the solenoid is de-energised.

Flow rates higher than the manufacturers limits will cause malfunction.

The electrical supply is crucial to both function and life of solenoids.

WIRING

For mobile type operation i.e. using direct current solenoids, significant voltage drop can be experienced in wiring. Small diameter wires running over long distances can cause serious loss of voltage at the solenoid. Also where several solenoids are operated together and the earth wires are collected to a common earth return high voltage losses can occur in the earth return.

The use of Marine Grade tinned cables and terminals can reduce the possibility of corrosion and consequent failure of the electric system where systems are used in corrosive atmospheres [such as marine].

TWO STAGE SOLENOID OPERATED VALVES

Single stage valves have functional and pressure drop limitations. When flows are high two stage valves are used.

This is a combination of a direct solenoid valve and a spool type mainstage which uses a large diameter spool. The body has larger flow passages etc.

To function the solenoid valve has to have pilot pressure available at all times of approximately 7 bar.

Operation of the solenoids directs fluid to either end of the main spool to shift it into the operating position. When this pilot pressure is released the main spool is returned to the neutral position by the return springs.

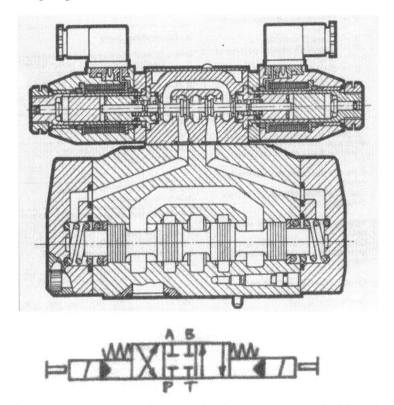

Illustration 58. 2 stage directional valve (spring centred) (Eaton)

WHAT CAN GO WRONG

Debris in the fluid can cause the spools to stick and the bores/spools to wear.
Springs can fail.

Each type of spool has its own functional limits. The flow forces on the spool as it is operating are a function of pressure and flow.

Any of the problems associated with direct operated valves can be experienced.

LEVER CONTROLLED DIRECTIONAL VALVES

Used in many industries particularly mobile and construction, the spool is connected to a lever which is used to move the spool into the operating position.

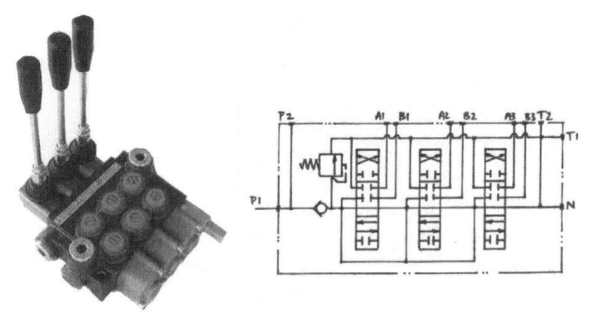

Illustration 59. Mobile directional valve (Hypro)

The symbol is typical for this type of valve (Hypro)

Lever operated valves can also be operated by remote control cables.

To keep leakage past the lands to a minimum for load holding purposes the spool/body diametral clearance is kept to a minimum by accurate machining and selection of the components.

Spools are often matched to the specific application by machining the spools i.e. providing slots, notches etc. to enable the operator to accelerate and decelerate actuators smoothly.

A fourth position can be provided often to give a float function to the cylinder i.e. both service ports connected to tank which allows the machine element to "float" or move freely as other parts of the machine are working.

PROPORTIONAL MOBILE VALVES

When combined with pressure control valves and shuttle valves to provide load sensing a fully proportional system can be created using either lever or electrically operated valves.

WHAT CAN GO WRONG

Wear can take place between spool and body if the fluid contains debris. The seal between body and spool may leak due to wear or damage.

Whilst simple in concept, high quality materials and machining ensure a long working life. I know of valves which are still operating satisfactorily after 30 years.

PROPORTIONAL DIRECTIONAL VALVES

Of similar construction to other spool type valves, the proportional valve solenoid moves the spool which has metering slots machined into it against springs by controlling the current to the solenoid. The spool takes up an intermediate position in the body and has directional and metering effect on the fluid the spool opening being proportional to the current applied to the solenoid. The first movement of the spool is to select the direction of flow i.e. connect the pressure port to a service port and a service port to tank. As the current is increased the flow through the valve increases.

The spools are machined to provide a linear increase in flow area relative to the linear spool movement after the first movement.

The solenoids have shaped pole faces and armatures which provide an output force which increases proportionally to the current applied to the coil regardless of the axial armature position relative to the pole face.

Sounds simple, but the components used are very precise.

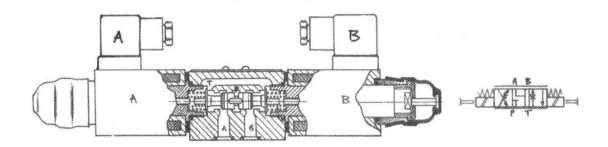

Illustration 60. Proportional directional valve

The variable current electronic controllers used need to have facilities to provide an initial step in current to move the spool into its operating position then increasing current to the full current capability of the solenoid.

By controlling the current rather than the voltage the affect of changes of coil temperature on coil resistance is eliminated.

Friction between spool and body and between solenoid tube and armature is overcome by using anti friction bearings and applying a dither component to the control current i.e. an alternating current on top of the DC.

Friction between spool and body due to trapped debris will result in lack of control.

A reduction of input voltage to the controller can result in loss of full spool movement due to inability to obtain full output current.

TWO STAGE PROPORTIONAL VALVES

The main spool is controlled by springs as in a single stage valve but the control current to the pilot valve solenoid is converted into a variable pressure by the pilot stage so that the variable pressure moves the main spool into position against the return spring or springs.

The spools are machined with metering lands to obtain a linear increase in flow area relative to spool linear movement.

The supply pressure to the pilot valve must always be kept higher than a minimum value to obtain valve function.

Friction between spools and bodies can be caused by debris. This will result in lack of control.

LOAD SENSING WITH PROPORTIONAL VALVES

Load sensing measures the pressure drop across a proportional valve when metering flow.

Inlet and outlet pressures are sensed by a control element which bypasses fluid to tank or destrokes a supply pump at a preset pressure drop i.e. a control element may be set to bypass flow at a pressure drop of for instance 15 bar. This means that the difference in pressure across the metering spool is always 15 bar regardless of load pressure and when the inlet pressure is over 15 bar higher than the load pressure excess fluid is bypassed to tank or the pump destroked to maintain the pressure difference.

Keeping the pressure drop to a constant figure enables the proportional valve to control flow by metering area exposed only i.e. if the area exposed increases in a linear manner relative to spool stroke the increase in flow will be linear.

A number of proportional valves can be used in conjunction with one control element by connecting the pilot pressure connections through shuttle valves or check valves.

Load sensing is an efficient way of controlling flow since the only pressure drop is that set by the control element and much less waste (heat) is generated than those systems which meter at relief or pump pressure.

SERVO VALVES

Essentially 4 port directional valves servo valves have spools and bodies (sleeves) machined to provide maximum overlap or slight underlap in the neutral position.

Operation of the pilot stage results in spool movement and therefore opening to allow flow into and out of an actuator. The spool position is sensed and the feedback used to ensure that the signal is matched when the spool is in the correct position.

Electronic feedback of the actuator position is normal, forming a closed loop system and providing very sensitive position control of an actuator.

Servo valves have been used extensively in aircraft for flight control positioning and other functions and are used to control flight simulators.

The pressure drops can be high so there may be limitations in controlling heavy loads.

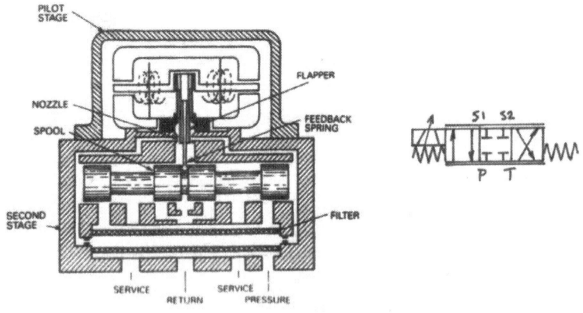

Illustration 61. Servo valve (Ultra)

WHAT CAN GO WRONG

Debris in the fluid cannot be tolerated. Fairly sophisticated electronic systems are needed to be effective.

TRANSMISSION VALVES

Size 3 solenoid valves have been in use by gearbox manufacturers for many years to control the operation of hydraulic clutches in heavy duty gearboxes for off road vehicles. They are however somewhat heavy handed as the clutch is either engaged or disengaged.

Recent developments are combined directional and pressure controls which provide rapid engagement but at a controlled rate. Operated by proportional solenoids these valves can be programmed to provide controlled engagement and disengagement to achieve efficient and smooth gear changes.

Patent No GB2325725 was assigned to the invention of a transmission valve filed by the author in 1997.

CHECK VALVES

Check valves/non return valves allow fluid flow in one direction only.

The simplest is a ball, seat and spring but this has limitations. The ball needs to be guided to avoid it damaging the seat as it reseats and also needs to be restricted in movement to avoid boxing the spring and thereby damaging it. For these reasons poppets which are guided to keep them in line with the seat and which incorporate a method of protecting the spring are generally used.

Return springs which offer resistance to valve opening of approx 0.5 bar are normally employed. Where check valves are used for other purposes i.e. for generating pilot pressure, springs which give a higher opening pressure can be fitted.

Debris in the fluid can cause leakage past the seat.

Selection of check valves is mainly dependent on the flow rate i.e. the pressure drop allowable at the flow through the valve.

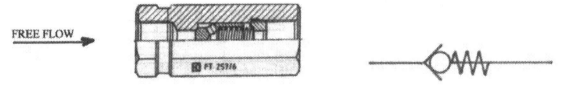

Illustration 62. Check valve (Isis)

SHUTTLE VALVES

Shuttle valves are 3 port valves which are used to connect the higher pressure from two sources to a common port the higher pressure having priority and at the same time isolating the lower pressure. These valves are used in load sensing circuits to connect the active service port of the proportional direction valve to the load sensing element or pump. They can be used individually or in cascades where a number of proportional valves are employed in the system.

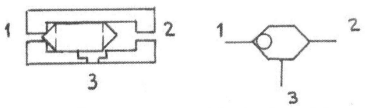

Illustration 62a Typical shuttle valve symbol

PILOT OPERATED CHECKS

Used to control accurately (without leakage) the position of cylinders, pilot operated checks consist of check valves which are unseated to allow flow in the check direction by the pressurization of a pilot piston.

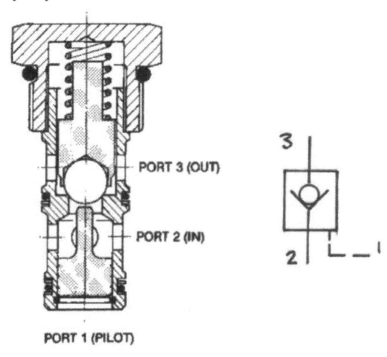

Illustration 63. Pilot operated check valve

Working out the ideal ratio between the check valve seat area and pilot piston area is not as straightforward as it seems as the pressure generated by the load and by the pilot pressure itself are significant factors particularly when the load is extending the cylinder.

The following calculation illustrates this:-

Care must be taken when applying pilot operated check valves to double acting cylinders to ensure that there is:

1. Sufficient pressure available to operate the pilot piston if there is a high pressure being held by the check valve.

2. Sufficient pressure to keep the check valve open after it is first opened if it has been holding against a high pressure.

These conditions can be met by using a counterbalance or flow control valve in the cylinder line which will generate enough pressure to keep the system stable under operating conditions. The following formulae enable the pressure settings to be worked out:

In static conditions ignoring friction the pressure generated in a cylinder

$$P_2 = \frac{L}{A_2}$$

Say the opening ratio of the valve is 3.5 to 1 i.e. the piston area is 3.5 times the seat area of the poppet. The additional pilot pressure to operate the pilot piston against the spring is so small that it can be discounted.

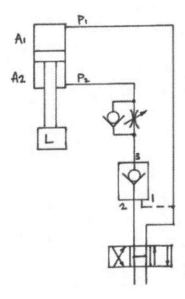

Illustration 63a. Pilot operated check circuit

The pilot pressure to open the check valve is therefore

$$P_1 = \frac{P_2}{\text{Opening ratio}}$$

1. $P_1 = \dfrac{P_2}{3.5}$

P_2 = minimum setting of restrictor valve

P_2 = $\dfrac{\text{static load L}}{\text{Area } A_2}$ + P_1 x ratio of cylinder areas

since any pressure applied at P_1 will increase the P_2 pressure.

2.   therefore P_2 = $\dfrac{L}{A_2}$ + $\dfrac{P_1 \text{ x } A_1}{A_2}$

Substituting 1 into 2

$$P_2 = \frac{L}{A_2} + \frac{P_2}{3.5} \text{ x } \frac{A_1}{A_2}$$

Therefore $P_2 - \dfrac{P_2}{3.5} \text{ x } \dfrac{A_1}{A_2} = \dfrac{L}{A_2}$

Simplifying this becomes

3. $P_2 = \dfrac{L}{A_2 - A_1/3.5}$

All the values in the right hand side are known so the minimum setting of the restrictor valve can be found. P_1 can also be found from equation 1 which is the minimum system pressure to open the pilot operated check valve and keep it open.

It can be seen that as the ratio of cylinder areas increase the P_2 pressure increases to the point where if the ratio of the cylinder areas was 3.5 it would be impossible to operate the system. On the other hand if the load on the cylinder was acting in the opposite direction, as the ratio of cylinder areas increases the P_2 pressure will decrease.

EXAMPLE

L = 100,000N
A_1 = 100cm^2
A_2 = 65cm^2

$$P_2 = \frac{100000}{65 - 100/3.5} = \frac{100000}{36.5}$$

$$= 2740 \text{N/cm}^3 = 274 \text{bar}$$

$$P_1 = \frac{274}{3.5} = 78 \text{bar}$$

To keep the check valve open the restrictor must be set to provide a pilot pressure of more than 78bar which in conjunction with the load generates a pressure in the rod end of the cylinder of 274bar plus.

In systems where there is a metering or pressure control between the check valve and tank a drain line is required to drain the annulus or top of the pilot piston to allow the pilot piston to function.

WHAT CAN GO WRONG

Debris in the fluid can cause damage and leakage past the check valve seat.

Without some form of metering control between the cylinder and the pilot operated check the valve will open then promptly reseat causing the jerky motion of the cylinder. (see previous example of calculation)

If a cylinder is moving a load with high inertia forces, very high pressures can be generated between the cylinder and the pilot operated check when the check closes enough to cause pipe and seal failure

LOAD CONTROL VALVES

The function is to stop actuators running out of control when subject to overrunning loads. They provide a pressure control valve which is opened in a controlled manner when the pilot piston is pressurized.

The modulating effect of the pilot pressure to the valve setting ensures a smooth function of the machine i.e. the higher the pilot pressure the more the valve opens.

The induced load will determine the valve setting and the pilot ratio will determine the pilot pressure.

Pilot pressure = valve setting - load induced pressure
pilot ratio

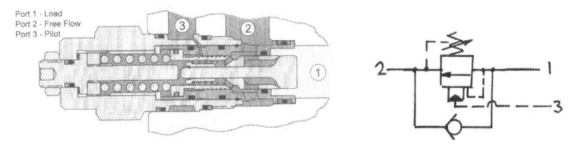

Illustration 64. Load control valve (Sun)

As the pressure builds up to the valve setting in port 1 the internal sleeve is moved up against the spring allowing flow from port 1 to port 2. Application of the pilot pressure assists this function. When the flow to the valve is reversed the central poppet check which is seated against the sleeve opens and allows free flow from port 2 to port 1.

The relief setting must be at least 30% above the theoretical load induced pressure to resist deceleration forces and to avoid leakage which will negate the load holding capability.

Setting the valve on the machine is difficult. Always set the valve on a test stand if possible.

Lower pilot ratios will give better control over cylinders. Higher pilot ratios will improve efficiency but possibly at the expense of stability.

High pilot ratios will normally provide adequate control of motors.

In some systems where there is high resistance downstream it is desirable to externally drain the pilot piston.

WHAT CAN GO WRONG

It is desirable to position the load control valve on or close to the actuator to eliminate failure of components hoses, seals etc. which may compromise safety and to avoid instability caused by sponginess in the system.

Where long pipes are used it is possible that the same volume of fluid can travel backwards and forwards through the load control continuously causing this volume of fluid to overheat. To avoid this it may be necessary to arrange the piping so that flow is generally in one direction only. This is especially important in continuously reciprocating actuators, agitators, paint strippers etc.

Debris in the fluid can cause damage and leakage past the relief valve seat.

FLOW CONTROL VALVES

Speed of actuators is often determined by the pump displacement. However, where a lower speed is required flow controls can be used. These can be as simple as an orifice.

Whilst an orifice appears to be straightforward if it is of any significant length the fluid viscosity can change the flow through it. Very short orifices are hardly affected at all by viscosity. The formula for flow through a short orifice does not have viscosity as a factor.

FLOW THROUGH AN ORIFICE

Assuming it is sharp edged, then the flow rate is proportional to the area x $\sqrt{\text{pressure drop}}$

Where the diameter is in inches and the pressure drop in p.s.i. Q = flow rate
$$Q = 94d^2\sqrt{\Delta P} = inches^3/sec$$

In metric terms
$$Q = 0.54d^2\sqrt{\Delta P} = litre/min$$

Where d = diameter in mm and ΔP is pressure drop in bar, Q = flow rate

These formulae assume an effective orifice area of 72% of the measured area. Because of the contraction of the effective area due to the dynamics of the flow, the longer the orifice is the smaller the effective area.

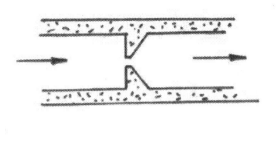

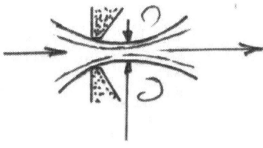

Illustration 65. Fluid flow through orifice, effective flow diameter is 85% of orifice diameter = 72% of area.

PRESSURE COMPENSATED FLOW CONTROLS

Flow controls start getting a little more sophisticated when it is required that the flow rate is independent of pressure drop. This type of valve is a pressure compensated flow control and has a compensator spool arranged to work either in series with the orifice or parallel to it.

When the compensator is in series the fluid flows through the control orifice. The pressure is sensed by the compensator from both sides of the orifice and restricts the flow to keep the pressure drop constant regardless of inlet or outlet pressures, and therefore provides a constant flow rate. Excess flow is dumped to tank via the relief valve or by destroking the pump.

Where precise flow rates are required flow controls with graduated dials can be used e.g. machine tools, moulding machines.

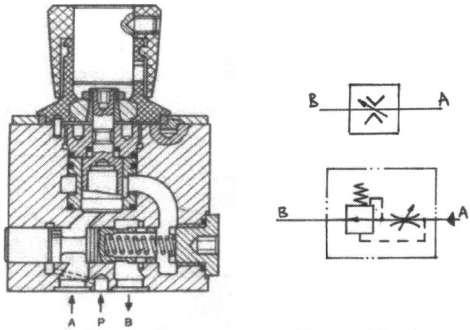

Illustration 66. Pressure compensated flow control (Eaton)

2 port valves Flow in the A port passes the adjustable orifice and then passes through the compensator spool and out of the B port. When the pressure drop across the orifice equals the spring force divided by the spool area, the spool moves across and restricts the flow.

The simplest form of pressure compensated orifice is where the orifice and compensator are incorporated into one part.

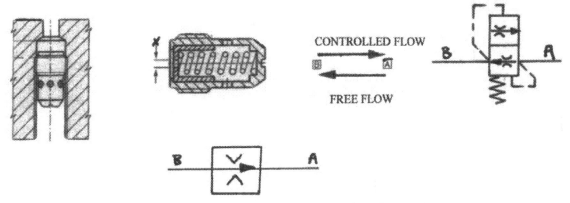

Illustration 67. Pressure compensated orifice (Isis)

As the flow passes from B to A the pressure drop across the orifice moves the spool until it restricts the flow by sliding over the holes in the sleeve. Different flow rates are obtained by a range of orifice sizes.

This device can be incorporated into pipework, circuit blocks etc. and is a cheap and practical device that works well.

3 port valves   When the compensator is in parallel the fluid flows through the orifice and when the pressure drop through the orifice is higher than the spring setting of the compensator the compensator opens and allows flow to tank or to secondary circuit i.e. excess flow is bypassed at slightly higher than load pressure, not relief valve pressure.

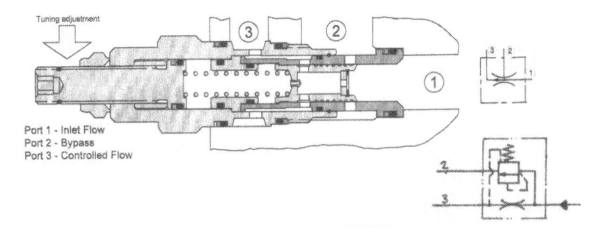

Illustration 68. Pressure compensated 3 port flow control (Sun)

Flow into port 1 passes the orifice and out of port 3. When the pressure drop equals the spring force divided by the sleeve area the sleeve moves to the left and allows excess flow to port 2 at the system pressure.

For multiple flow rates the use of one or more directional valves can be used to select a number of orifices all working in conjunction with the compensator. The orifice can be in the form of plates which fit in the seal recesses of the directional valve.

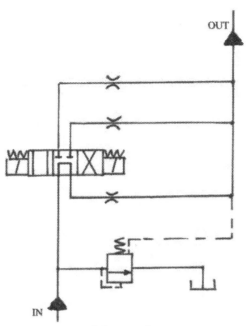

Illustration 69.   System using modulating element as compensator (Sun) to provide 3 flow rates

Where constant flow is required from a variable speed engine driven by a variable displacement pump the pressure drop across an orifice can be sensed to control the pump to a constant value e.g. vehicle refrigeration drives.

PROPORTIONAL FLOW CONTROLS

To meet the requirements of a high reach forklift truck manufacturer the author developed in 1997 a two port proportional flow control valve to control the flow to the lift cylinder and provide smooth, precise positioning. This valve incorporated an integral pressure compensation feature based on using flow forces so that the machine function was the same whether loaded or unloaded. Patent GB2326215 was assigned for this invention.

Conventional proportional flow controls are created by opening and closing the control orifice using an electric motor or proportional solenoid

WHAT CAN GO WRONG

Debris in the fluid can cause the compensator spool to stick.

Because the compensator spool has to travel a small amount before it gets into the operating position there can be an overshoot of flow when the system is started. This can be avoided to some degree by having a mechanical adjuster to keep the spool movement to a minimum.

All flow controls waste energy to some degree and therefore create heat.

FLOW DIVIDERS/COMBINERS

Where flow needs to be split to operate two systems simultaneously a spool type valve can be used. This can include the same feature when flow is reversed.

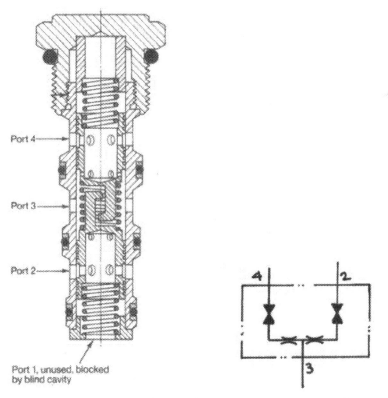

Illustration 70. Divider-combiner pressure compensated (Eaton)

Flow is applied to Port 3 and equal flows are produced at ports 2 and 4. When the flow is reversed flow is controlled equally from ports 2 and 4 to port 3.

POPPET VALVES

Basically a check valve with the facility for applying a pilot pressure to close the valve, poppet valves can be used in many systems to provide a single on/off function. For instance a 4 port directional valve function can be replicated using 4 or 5 poppet valves. Where flows are high they are particularly effective.

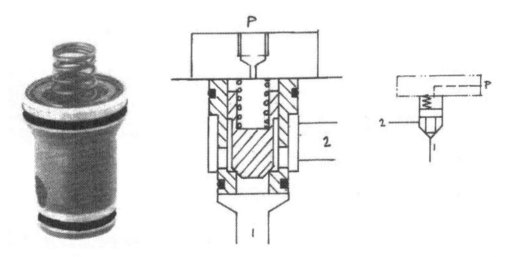

Illustration 71. Poppet type valve

Pressure can be applied to ports 1 or 2 and the valve will open or close depending on the application of pressure to the pilot port.

SOLENOID OPERATED POPPET VALVES

Normally 2 or 3 port the solenoid is energised to either open or close the valve. Leaktight operation is obtained by using a seated poppet.

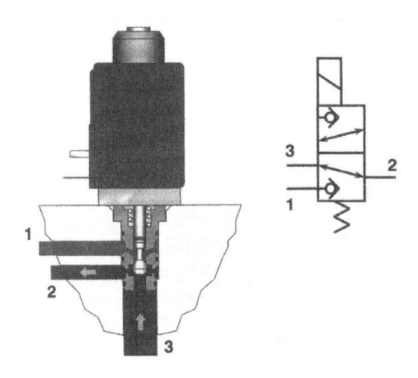

Illustration 72. Solenoid operated poppet valve (Eaton)

With the solenoid de-energised ports 2 and 3 are connected. With the solenoid energised 1 and 2 are connected.

WHAT CAN GO WRONG

Debris can lodge on or damage the seat. The quality of the machining of the seat and poppet has to be very accurate to achieve leakfree function and long life. The fit and resistance to movement of seals is a vital aspect of the design

CARTRIDGE VALVES / CIRCUIT BLOCKS

Many valve functional elements can be incorporated into a cartridge form either slip in or screw in. The advantages are clear. The cartridge components can be made on high volume production machinery very accurately at relatively low cost.

Any combination of valve function can be built into a circuit by using a manifold or circuit block either in steel or aluminium. These blocks can also be machined accurately and economically using CNC machine tools even in quite small batches.

Cartridges and circuit blocks have been a major success story at the end of the 20th century.

Design of blocks is now assisted by the use of CAD programmes.

IDENTIFICATION OF PORTS/VALVES

It is a requirement of block system suppliers to identify ports and valves relative to the circuit diagram.

Illustration 73. Typical circuit block

INTERCHANGEABILITY – CARTRIDGE VALVES

During the development of cartridges manufacturers went in different directions with the result that several "standards" have developed. This has a limited effect on system integrity because there generally is no problem obtaining spares and these are rarely required anyway.

Major manufacturers who developed their own cavities were Modular Controls, Fluid Controls, Sterling and Sun. It does mean that independent block manufacturers have had to tool up with a multiplicity of tools to produce these different cavities. Each kit of tools generally consists of boring, finishing and tapping tools.

In recent years an international standard has been agreed for screw in cartridges based on using metric threads which is gradually obtaining acceptance. (ISO 7789)

WHAT CAN GO WRONG

When used outdoors or in hostile environments cartridges and blocks must be protected against corrosion.

Screw in cartridges can be compressed radially in the thread area by the action of tightening the cartridges into the block. Cartridges which include close fitting spools may be affected by this and the spools may get pinched causing malfunction.

Always tighten cartridges to the manufacturers recommended torque and do not exceed it. This torque may be substantially less than the thread size would suggest as the forces acting to push the cartridge out are relatively low compared to the thread size.

Circuit blocks by definition have many interconnecting holes in them some of which must be plugged.

For small internal and external plugs and orifices the author's preference is to use plugs with taper threads. There is one internationally recognized taper thread which is the American NPTF and both tooling and plugs are readily available.

High strength materials and good surface finishes need to be used particularly in aluminium where it is easy to pull threads out. De-burring of interconnecting holes needs to be done meticulously to avoid swarf or shards of metal becoming loose during system operation and jamming valves and other components.

Cleaning after manufacturing must be thorough and meticulous inspection of internal passages to avoid these problems must be part of the manufacturing process.

BLOCK MATERIALS

For pressures up to 250 bar aluminium cast iron or steel is used. For higher pressures steel is generally specified because of the higher strength. Aluminium is usually 6082 although where blocks are used in automotive power packs cast aluminium is sometimes used.

The Author's preference is to finish the aluminium blocks by Anodising, to provide corrosion resistance and resistance to "galling" by threaded fasteners.

Block material needs to be defect free. Porosity leads to leakage and extrusion faults (usually at the end of the bar) can cause leakage and compromise material strength.

For external plugs a wide variety are in use:

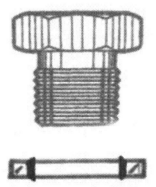

Illustration 74. BSP/Metric plug with bonded seal

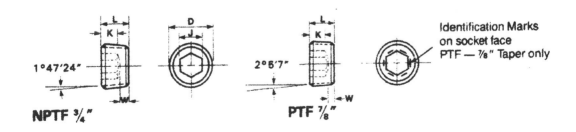

Illustration 75. NPTF taper plugs with internal hexagon fixing

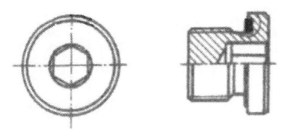

Illustration 76. Metric/BSP plug with trapped seal

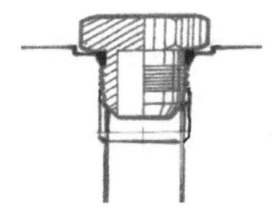

Illustration 77. SAE unified thread plug with trapped 'O' ring

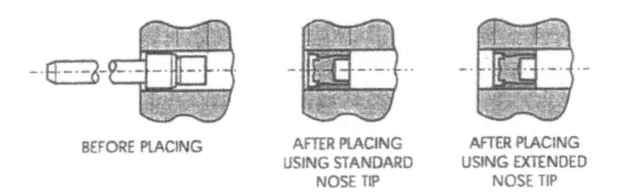

BEFORE PLACING

AFTER PLACING
USING STANDARD
NOSE TIP

AFTER PLACING
USING EXTENDED
NOSE TIP

Illustration 78. Pop rivet plugs set by a tapered rivet (Textron)

The requirements of all plugs is that the machined component are to a high standard i.e. threads, counterbores, seal faces, bore diameters, surface finishes etc.

NPTF threads require the addition of adhesive sealant for external plugs which may require the application of heat to enable removal.

The female thread has a taper of 3⁄4 inch per foot and there are two types of plug one which has a thread of 3⁄4 inch per foot (NPTF) and the other type has a taper of 7⁄8 inch per foot (NPT). The 7⁄8 taper plug is generally shorter in length and fits deeper into the thread to finish subflush.

For these reasons NPTF threads have to be formed by tapping to a specific depth tolerance to suit the type of plug being used.

Pop rivet plugs and Cup type plugs have pressure limits many times higher than system pressures. Machining tolerances and surface finishes must conform to the manufacturers specifications.

Adequate material must be provided around all holes to support the radial loads induced by the plugs.

VALVE MOUNTING SURFACES (SUBPLATE SURFACES)

A range of valve mountings for all types of valves is in general use worldwide and is well recognised. However there are differences in performance and details between different manufacturers valves so great care in comparing all aspects of a valve's performance, sizes, finishes etc. needs to be taken before changing one manufacturers valve with another.

The standards control the maximum width of a valve but not the length or the height. The standards also specify the flatness of the surface.

The mounting principle is that the mounting surface is flat and has the ports in the form of holes. The ports are sealed with seals mounted either in a plate or in recesses in the valve body. The valve is fixed down with either 4 or 6 high tensile cap screws tightened to a specific torque, Dowels or unequally spaced bolt positions ensure that the valve cannot be fitted the wrong way round.

The international adoption of the size 3 valve mounting surface in the 1970s resulted in massive growth in quantities of these valves worldwide. They have a maximum width of 50 mm. The smaller size 2 40 mm wide surface adopted in the 1990s has had less success mainly due to the physical restraints imposed by the mounting surface design. The opportunity to introduce a design which could have been equally as successful as the size 3 was missed despite the authors efforts to get one adopted.

WHAT CAN GO WRONG

Seals can be extruded if cap screws or studs are not tightened to the correct torque.

Mounting surfaces need to be flat within 0.01mm/100mm and the surface finish better than 0.8µm

VALVE MODULES

The major advantages of subplate mounted valves is that one or more control modules can be fitted between the directional valve and the subplate creating circuits by building up the valve stack eliminating a lot of pipework and labour and making servicing much easier. Module heights are standardised (size 3 = 40mm) but there are variations from manufacturers which mean bolts or studs of intermediate lengths are required.

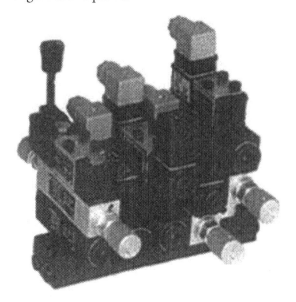

Illustration 79. Size 3 manifold with valve stacks

WHAT CAN GO WRONG

Great care must be taken in replacing one manufacturers valve with another.

There are standards for module thickness but length can vary and valve function may not be identical.

Despite the use of dowels and offset fixing hole positions it is not impossible to assemble the modules incorrectly particularly the size 3.

15

SEALS

N.A.Peppiatt - Hallite Seals International Ltd

Seals are a vital element in any fluid power system, but their importance in the satisfactory performance of equipment often does not receive the attention it deserves. This chapter aims to provide greater understanding of seal behaviour by reviewing typical seals found in fluid power applications and highlighting many of the major problem areas and pitfalls to be avoided. Most commonly, a polymeric component is to be found at the heart of the sealing system. The other components include the housings and the fluid media to be sealed, and in the case of dynamic seals, the bearings that guide the moving parts. It is necessary to begin with the caveat that if there is a problem with one of the other components, then the polymeric seal is the first part to suffer.

Seals can be divided into two types, static and dynamic. The commonest seal profile is the toroidal ring or O-ring . This is typically, but not exclusively, used as a static sealing element. An examination of the major seal manufacturers' catalogues will show that a bewildering variety of other profiles have been developed for dynamic applications both for reciprocating and rotary applications.

O-RINGS

O-rings are generally selected from standard sizes that cover the vast majority of applications e.g. BS1806 for inch sizes and BS4518 for metric sizes. Unfortunately, ISO3601, which is a hybrid of metric shaft and bore sizes and inch cross section O-rings, is not yet fully developed and cannot be recommended at the time of writing. This ISO standard has a fundamental problem in that the O-ring cross-sections are based on the inch ones of BS1806, which are, of course, designed for nominal fractional inch housing cross-sections, i.e. 1/16", 3/32", 1/8", 3/16" and 1/4". As a result the housing cross-sections do not in general relate to convenient metric (mm) cross-sections and a given O-ring does not suit both a convenient bore size and rod size. This has resulted in a proliferation of O-ring sizes in order to fit a reasonable range of housing sizes and possible confusion of sizes with the existing well established inch (BS1806) range. A much fuller explanation can be found in reference (1), with an update on developments in reference (2).

The O-ring functions as a seal by being compressed in its housing as shown in Fig 1. This squeeze generates an initial sealing force that prevents fluid passing as pressure is applied. As the pressure is increased the sealing force increases and as the pressure is reduced the sealing force reduces. At all times the stress at the sealing contact remains slightly higher than the applied fluid pressure, providing an automatic self-compensating system. The seal is distorted in the housing under increasing pressure until it is squeezed against the housing surfaces away from the pressure after which it behaves like an elastic fluid.

When a seal fails to hold pressure it is because of the loss of the initial sealing force.

The O-ring is an example of a positive squeeze seal. When installed as shown in Fig 1 it will seal if the pressure is applied from either direction.

Rubbers make the ideal materials for O-rings and similar seals as they can deflect elastically at high strains, are soft enough to conform to metal housings and have high bulk moduli. The surface finish of the housings is important to prevent fluid from seeping past the seal and to minimise its transfer caused by seal movement as it is pressurised. A maximum roughness of 1.6μm Ra is generally recommended for static sealing faces.

MATERIALS

A variety of polymeric materials, both rubber and plastic, are used in the construction of seals. Nitrile rubber (a copolymer of acrylonitrile and butadiene rubbers) is the most common compound found in fluid power seals, because of its oil resistance and relatively low cost. It is widely used for O-rings and can be reinforced by textiles such as cotton. This composite has provided the basis of many reciprocating seal designs, but has now been largely superseded in this application by thermoplastic elastomers such as polyurethane. The harder plastics, such as acetal or nylon, are used to contain the seal material at higher pressures and as bearing elements. Table 1, reproduced from BFPA/P81, lists a wide variety of materials found in fluid power seal and bearing construction and gives an indication of their suitability for use in fluid power and some other media. Such recommendations can only be regarded as a guideline. As a result of the different additives in proprietary seal compounds and proprietary fluids their compatibility is best confirmed by long-term tests or confirmation from field experience.

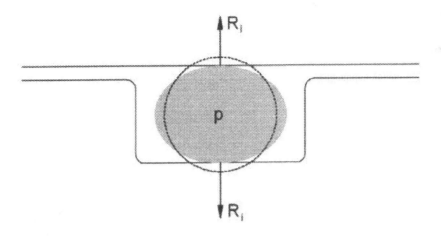

Squeezing O-ring in housing results in initial sealing force R_i and internal stress p

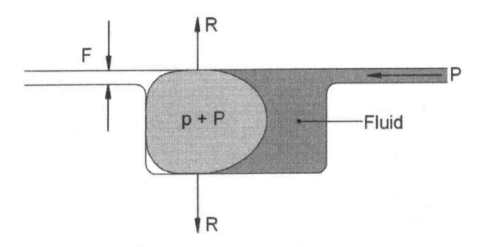

Applied pressure P adds to the internal stress p and increases sealing force R.
Seal can extrude through gap F.

Fig (1) Sealing mechanism of O-Ring

Material (1)	Continuous material service temperature range °C	Intermittent material service temperature range °C	Service Fluids (3)																				
			Fluids based on mineral oils				Greases		Fuels			Fire-resistant hydraulic fluids					Environmentally acceptable hydraulic fluids (5)				Other service fluids		
			Motor oils	Hypoid gear oils	Automatic Transmission fluid	Hydraulic oils ISO 6743-4 (HL, HM, HV)	Mineral oil based greases	Silicone based greases	Diesel fuel	Fuel for gasoline/petrol engines - normal	Fuel for gasoline/petrol engines - super	HFA fluids ISO 6743-4 (5/95 Water based)	HFA fluids ISO 6743-4 (60/40 Invert emulsion)	HFB fluids ISO 6743-4 (Water/glycol)	ALKYL (aero) (Phosphate ester) ISO 6743-4 HFDR fluids	ALKYL (ind.) (Phosphate ester) ISO 6743-4 HFDR fluids	Vegetable oil based HETG	Synthetic ester based HEES	Polyalkyl glycol based HEPG	Synthetic hydrocarbons HEPR	Water	Air	Brake fluids
Temp range for fluid °C			150 to -40	150 to -40	160 to -50	100 to -30	100 to -30	250 to -50				60 to 5	60 to 5	60 to -30	100 to -50	150 to 0	60 to -10	100 to -40	100 to -50	150 to -50	60 to 5 (4)	200 to 2	130 to -50
								Maximum continuous service temperature in fluid °C (2)																
NBR 70 IRHD / NBR 90 IRHD Nitrile (medium)	100 / -30	120 / -30	100	90	100	100	100	100	*	*	*	60	60	60	NS	NS	60	60	60	100	80	100	NS
FKM 70 IRHD / FKM 90 IRHD Fluoro-elastomer	200 / -20	250 / -20	150	150	160	100	100	200	150	150	150	60	60	NS	NS	150	60	100	80	150	100	200	NS
EPDM 70 IRHD / EPDM 80 IRHD	120 / -50	150 / -50	NS	NS	NS	NS	NS	120	NS	NS	NS	NS	NS	60	80	80	NS	NS	NS	NS	120	120	120
Peroxide cured VMQ 70 IRHD Silicone	200 / -55	250 / -55	*	*	*	*	100	*	*	*	*	NS	40*	40*	NS	NS	60*	NS	NS	*	100	200	80
HNBR 75 IRHD Hydrogenated nitrile	130 / -30	150 / -30	130	110	130	100	100	130	NS	NS	NS	60	60	60	NS	NS	60	60	80	130	130	130	NS
IIR Butyl	120 / -40	140 / -40	NS	NS	NS	NS	NS	120	NS	NS	NS	60	60	60	100	120	NS	NS	NS	NS	120	120	80
FFKM Perfluoro-elastomer	300 / -20	200 / -20	150	150	160	100	100	200	150	150	150	60	60	60	100	150	60	100	100	150	150	200	130
Polyester pu (AU)	100 / -40	120 / -40	100	100	100	100	100	100	60	60	60	40*	40*	40*	NS	NS	60	60	50	100	40*	40	NS
Polyether pu (EU)	100 / -40	110 / -40	100	100	100	100	100	100	60	60	60	60*	60*	NS	NS	NS	60	80	60	100	60*	80	NS
Polyester elastomer	100 / -40	110 / -40	100	100	100	100	100	100	60	100	100	60	60	60	100	100	60	80	60	100	60	80	80
Polyamide (PA)	100 / -40	120 / -40	100	100	100	100	100	100	100	100	100	60	60	60	100	100	60	100	100	100	60	80	80
Acetal (POM)	100 / -40	120 / -40	100	100	100	100	100	100	100	150	100	60	60	60	100	100	60	100	100	100	80	80	80
Polyphenylene sulphide (PPS)	200 / -40	200 / -40	150	150	160	100	100	200	150	150	150	60	60	60	100	150	60	100	100	150	150	200	130
PTFE	200 / -200	200 / -200	150	150	160	100	100	250	150	150	150	60	60	100	100	150	60	100	100	150	150	200	130
Thermosetting Polyester resin	100 / -50	130 / -200	100	100	100	100	100	100	100	100	100	60	40	60	100	60	60	100	100	100	80	100	NS
PEEK	250 / -65	300 / -65	150	150	160	100	100	250	150	150	150	60	60	60	100	150	60	100	100	100	150	200	130

* Denotes values vary greatly for individual elastomers within this group

NS – Denotes that the elastomer is not suitable

Table 1 showing the typical maximum continuous working temperatures (°C) and temperature ranges for seal materials in fluid power applications.

TEMPERATURE RANGE OF SEAL MATERIALS

Table 1 gives guidelines for both the temperature range of seal materials and for fluid media. For the seal the upper temperature is one above which unacceptable permanent set or degradation of the seal polymer occurs. The glass transition temperature of the polymer governs the lower limit. This is the temperature below which the elastomer is no longer flexible or elastic but is in a hard glassy state. The service temperature of the harder thermoplastics e.g. acetal or nylon is below their glass transition temperature.

In the nitrile copolymer, the oil resistance and upper temperature limit is increased by increasing the acrylonitrile content, but this also raises the glass transition temperature. A low acrylonitrile content is required for low temperature flexibility and service. Nitrile rubbers are generally categorised in three classes (Reference 3);-

High Nitrile (Acrylonitrite content 35% to 45%)
Resistance to mineral based fluids is excellent and to hydrocarbon fuels is good. Strength, resilience, abrasion and high temperature resistance are reasonable but performance at low temperature is poor.

This grade is used primarily in contact with aromatic fuels and mineral oils.

Medium Nitrile (Acrylonitrile content 27% to 34%)
Resistance to mineral based fluids is excellent, although resistance to fuels is not usually good enough. Strength, resilience, abrasion and heat resistance are reasonable. Low temperature resistance is adequate for most applications but weathering or ozone contact will result in crazing and cracking, particularly under tensile stress or flexure. This grade is the most commonly used of all polymers in hydraulic sealing. The temperature range in Table 1 refers to medium nitrile rubber.

Low Nitrile (Acrylonitrile content 18% to 25%)
It has some resistance to mineral based fluids but the main advantage is its low temperature capability. Strength, resilience and abrasion resistance are reasonable.

It is important to remember that all nitriles are non-resistant to castor oil and castor based vehicle brake fluids and should never be used in applications employing these fluids.

HARDNESS OF SEAL MATERIALS

The hardness of a seal material determines its resistance to extrusion under pressure through narrow gaps, clearances etc. Manufacturers generally offer three levels 70, 80 and 90 IRHD.

Fig (2) shows the relations between hardness and clearance for static seals. Dynamic seals would use anti-extrusion back up rings for pressures over 100 bar.

Anti-extrusion rings are required when considering conditions to the right of the relevant hardness curve.

For temperatures above 100 degrees C use the curve for the next harder rubber

Materials below the nominal 70 IRHD are excluded from this graph as being considered impractical for pressure above 50 bar.

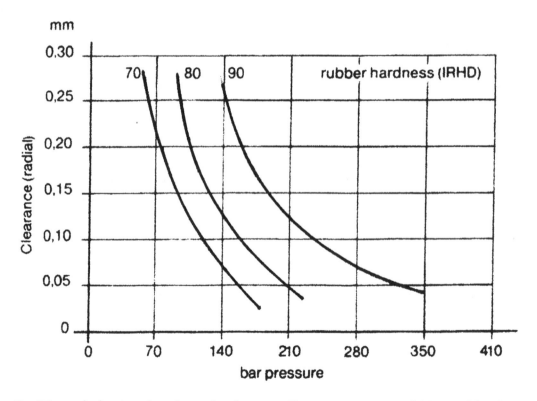

Fig (2) graph showing the relationship between Clearance pressure and Material hardness.
Reference (3)

DYNAMIC SEALS

There are three basic forms:

- Reciprocating seals

- Low pressure rotating shaft seals (rotary lip seals)

- High pressure rotating seals (mechanical seals)

The mechanics of these seal types are markedly different

RECIPROCATING SEALS

Apart from static seals, these are the commonest seals found in hydraulic equipment. They seal the rods and bores in cylinders and valves (both hydraulic and pneumatic). Fig 3 is a cross-section of a typical double acting hydraulic cylinder showing the positions of the major seals.

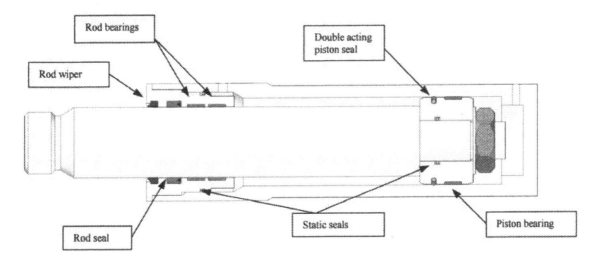

Fig (3) a cross section through hydraulic cylinder showing seals.

The piston seal is double acting in that it can hold pressure from both the annulus and full bore. The rod or gland seal is single acting and seals the opposite end of the annulus to the piston seal. The wiper prevents dirt from the outside entering the gland area, which would otherwise cause fluid contamination and damage to the gland bearing and seal. The sealing performance requirement of the gland sealing system - rod seal(s), bearings and wiper - is stringent because of the market demand for leak-free hydraulics.

The dynamic seals shown are those that might be used typically in a contemporary light/medium duty general-purpose double acting cylinder. The rod seal is a polyurethane U-ring. Unlike the O-ring the U-ring can be pressure energised in one direction only. The double acting piston seal shown consists of a thermoplastic elastomer face, which is energised by an O-ring. The wiper is also likely to be of a thermoplastic elastomer material. Polyurethanes and other thermoplastic elastomers are now commonly used for dynamic sealing elements because of their superior wear resistance compared with nitrile rubber. A typical pneumatic cylinder sealing arrangement is shown in Fig 4. Here thin lip U-rings from nitrile or polyurethane are typically used to minimise seal friction.

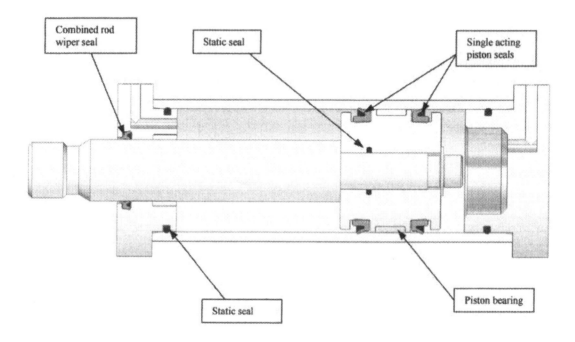

Fig (4) a cross section through Pneumatic cylinder showing seals

All the seals described above are squeezed radially, as the O-ring described earlier, and require an axial clearance in the housing. The result of this is that for each seal there are two sets of dimensions: - the actual seal dimensions and the housing dimensions as shown in Fig 5. The overall seal length will be less than the housing length and the maximum seal section will be greater than the housing section.

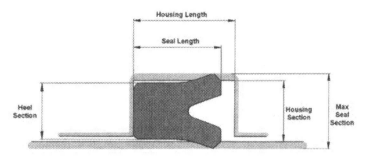

Fig (5) critical dimensions of a rod seal

WHY DOES A SEAL FAIL TO HOLD PRESSURE

This is a result of the loss of the initial sealing force. Generally, this is caused by one or more of the following factors: -

- Extrusion
- Wear
- Compression set
- Shrinkage (fluid compatibility)
- Dieseling damage
- Damage on assembly

EXTRUSION

As indicated earlier sealing polymers are necessarily relatively soft materials so that they can conform and seal against the hard metal counterfaces. Gaps must exist in housings, even if these are only location or bearing clearances. If sufficiently high, the pressure will force the sealing polymer down these gaps causing extrusion damage. Examples of seal extrusion failure are given in Fig 6, which shows examples of typical O-ring extrusion and a thermoplastic elastomer faced double acting piston seal where the face has extruded as a result of the diametrical gap between the bore and the piston being too large. In the case of O-rings, back-up rings (often PTFE) are recommended for pressures above 100bar to prevent this phenomenon. In dynamic seals hard plastic, e.g. acetal, anti-extrusion rings increase the resistance of the seal assembly to extrusion.

The extrusion gap is a result of housing tolerances, running clearances, bearing tolerances, ovalities, tube dilations and distortions caused by side loads.

See Fig (2) for the relationship between maximum clearance, pressure and seal hardness.

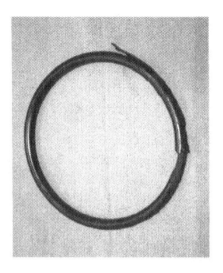 

Fig (6) typical examples of extrusion failure: Left – O Ring; Right – Double acting piston seal (mounted on piston)

WEAR

Wear can result if two surfaces rub together. The wear rate of a polymeric seal is affected by such factors as: -

- Surface finish of the dynamic surface against which the seal rubs
- Surface speed
- Temperature
- Fluid media
- Contamination

Two of the most important factors are the fluid media and surface finish.

FLUID MEDIA

Hydraulic mineral oils, for example, provide good lubricity, but this cannot be said of high water based fluids (e.g. HFA) which are, nevertheless, used successfully in a number of major applications, including longwall mining. Reference (4).

Compressed air can be moist and transporting an oil vapour that will provide lubrication for seals or it can be dry and "non-lube", in this case the seals will rely on the initial lubrication on assembly.

DYNAMIC SURFACE FINISHES

Piston rods are generally hard chrome plated. This gives an excellent tribological surface and provided the rods are produced by an established supplier within a surface finish range of 0.1 to 0.4µm Ra no major problem should ensue, although the optimum surface finish may well depend on the seal material. Reference (3).

Bore surface finishes can be more problematic. The typical methods of obtaining a bore finish are summarised in Fig 7. Drawn over mandrel (DOM) tubing, as is, can be adequate, or a potential disaster depending on the actual surface texture achieved. Increasing use is being made of Special Smooth Inside Diameter (SSID) DOM tubing, but in certain circumstances, mainly when the seal is being driven into the pressure, it can lead to wear of the seal through flow erosion. Reference (6). Such tubing requires careful specification. Skived and roller burnished tubing is very smooth (less than 0.1µm Ra) and may be too smooth for rubber sealing elements in some applications. True honed tube, produced to between 0.1 and 0.4µm Ra is the most expensive finish, but it is the quality solution.

COMPRESSION SET

The initial sealing force will be lost if the seal material does not recover to near its initial state on removal from the housing. The feature of rubber-like materials (elastomers) is their ability to store energy and their elastic recovery. However, they can be permanently deformed, in particular through the application of excessive heat. A comparative measure of the propensity of a seal material to take permanent set is the compression set test (e.g. BS903 Pt A6)

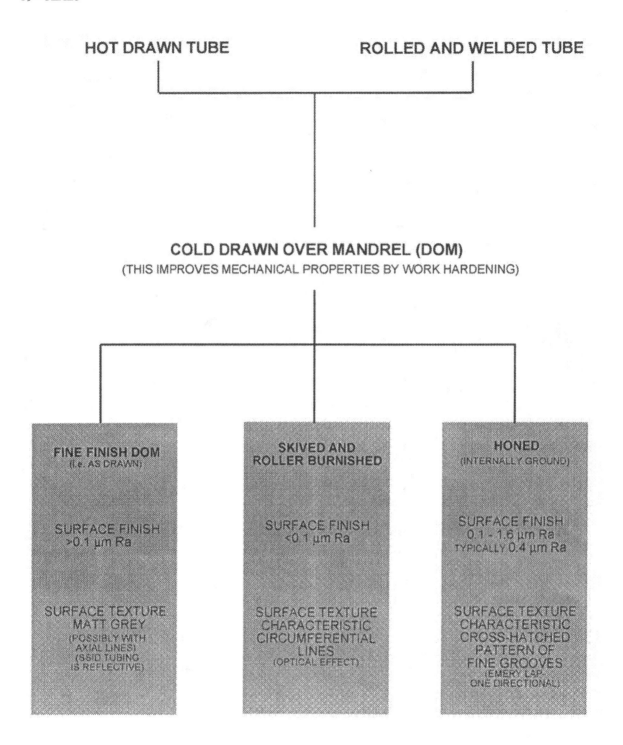

Fig (7) Methods of manufacturing of tubes for hydraulic cylinders and resulting surface textures.

FLUID COMPATIBILITY

Fluid media can cause a polymer to shrink or swell. Some slight swell is not generally a problem providing the seal housing does not over fill leading to accelerated wear or extrusion. Shrinkage can lead to loss of the initial sealing force. An example is the leaching of plasticisers out of low temperature nitrile formulations. A further compatibility effect worth noting because of the increased use of polyurethane seals is the effect of hydrolysis, or chemical breakdown of the material caused by water, which can occur quite rapidly in some formulations. For these reasons the compatibility index of ISO6072 takes into account change of hardness, change of volume, change of tensile strength and change in elongation at break, after the immersion period.

DIESELING DAMAGE

Seals can suffer "dieseling" damage or more correctly damage by compression ignition explosions. If air bubbles exist in the hydraulic oil, and if the cylinder is in such a position as these form by the seal, and the fluid experiences a rapid pressure rise, detonations of the oil/air mixture can occur which damage the seal. This often leaves a charred appearance on the damaged seal.

ASSEMBLY DAMAGE

It is not unusual for seal failure to result from assembly damage. Common causes are:

- Inadequate lead-in chamfers
- Lack of lubrication on assembly
- Passing unprotected or sharp ports in cylinder walls
- Contact with threads, burrs and sharp edges
- Contamination with swarf or other debris
- Misalignment on assembly
- Sharp assembly tools

Refer to seal suppliers for assistance in overcoming these problems.

LEAKAGE AND FRICTION

As a result of the understandable demands for leak-free hydraulics the dynamic leakage performance of the single acting gland sealing system is critical, and modern seal profiles are capable of giving minimal leakage. The problem is that the reciprocating seal works in boundary lubrication or at best a mixed lubrication regime, requiring the boundary lubricant to prevent wear. A single lip wiper working with a seal will often give increased leakage compared to the same seal without a wiper because the sharp wiping lip tends to prevent the return of the boundary lubricating fluid, and causes it to accumulate outside the gland. Reference (7). This affect is often overcome by the use of a double lip wiper or other secondary seal. The danger here is that if the main seal is a positive squeeze profile or for some other reason the main seal is unable to vent interseal pressure build up can occur. Reference (8). This trapped pressure can blow the wiper out of its groove or otherwise damage the main seal.

It is for this reason, that positive squeeze piston seals should not be used back to back for a double acting application. It is this pressure-trapping problem which led to the design of true double acting seals. It should be noted that the dynamic leakage of a double acting piston seal is far less critical than that of a single acting seal, where a run of oil from the rod is immediately visible. On the other hand the transfer of a small amount of oil from the annulus chamber to the full bore of a cylinder or vice-versa is not generally important, whereas the ability of the piston seal to provide a good static seal to hold position often is.

Friction of reciprocating seals can be inconsistent and difficult to predict, again as a result of the lubrication regime in which they operate. This is not, in general, a critical problem as the power required in overcoming that seal friction is a very small percentage of the power available in a hydraulic cylinder. In low pressure pneumatic cylinders seals with thin flexible lips are used to minimise the sealing force and hence the friction forces. Friction often only becomes significant when the "stick-slip" phenomenon occurs. This is a result of the static breakout friction being higher than the running friction and is particularly common at very slow speeds. The seal will deflect until there is sufficient force to overcome the break-out friction at the contact and then it will slip until the friction is greater than the pulling force resulting in a juddering motion that can excite natural frequencies in the cylinder or associated equipment. This problem is far less prevalent with the modern compact seal designs as shown in the cylinder of Fig 3 and can be virtually eliminated by the use of seals with PTFE dynamic seal faces. However, it should be pointed out that this latter type of seal does have other limitations. The most suitable seals for an application are usually found by a judicious mixture of experience and testing, both in the laboratory and in the field.

LOW PRESSURE ROTATING SHAFT SEALS

Used extensively throughout mechanical engineering rotary shaft seals retain fluid in housings where a shaft protrudes, they are sometimes known as rotary lip seals.

Comprising a rubber moulding incorporating a metal insert to stiffen it, the seal is an interference fit in the housing and the rubber is formed into a lip which is energised by a tension spring against the shaft fig (8) shows a typical arrangement. The same quality of materials are used to make the seals as the static and dynamic seals described earlier in this chapter. The shaft hardness where the lip is in contact is a factor in obtaining long life.

An external lip is often incorporated into the seal to prevent ingress of dust / water /dirt from outside the housing.

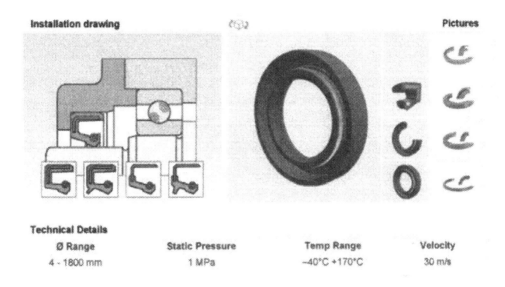

Installation drawing

Pictures

Technical Details

Ø Range	Static Pressure	Temp Range	Velocity
4 - 1800 mm	1 MPa	–40°C +170°C	30 m/s

Fig (8) Typical Rotary Seal, reference (3)

Pressures contained by rotary lip seals can be as high as 10 bar but in a dynamic situation the pressure applied needs to be as low as possible to obtain a long life.

HIGH PRESSURE ROTATING SEALS

 Generally comprising a bearing material such as carbon formed into a ring which presses against a metal face the high pressure seals allow pressures in the housings of motors and pumps to be as high as system pressures. Initially the sealing ring is held in contact with the face by a compression spring. Pressurising the housing of a motor or pump produces an outwards force on the shaft equal to the seal area times the pressure so axial needle roller bearings are used to contain the force.

STORAGE OF SEALS

Most polymeric items including vulcanised rubber, such as nitrile rubber, and other elastomers tend to change their properties during storage and may become unserviceable. This may be due to hardening, softening, cracking, crazing or other degradation and may be the result of oxygen, ozone, light, (particularly ultra violet), heat and/or humidity. For these reasons batch identification and strict rotation of elastomeric seal stocks should be practiced.

The storage conditions should keep seal components away from ultraviolet light, ozone and excessive temperature and/or humidity in an undeformed state. Contact with liquids and other rubbers should be avoided. BS7714 gives comprehensive guidance on storage conditions and the recommended storage life of typical sealing materials.

REFERENCES

1. Flitney, Bob, "International O-ring standards: where are they?" Sealing Technology, October 2003
2. Flitney, Bob,"Standards update: O-ring standards progress mm by mm" Sealing Technology, August 2006
3. Trelleborg Dowty Seals, Technical Data Handbook.
4. Peppiatt, N.A., "Reciprocating seals for water based fluids". Paper presented at the 47th International Conference on Fluid Power, Chicago, USA 23-25th April 1996
5. Flitney, R.K. and Nau, B.S., "Effects of surface finish on reciprocating seal performance". Paper presented at the 14th International Conference on Fluid Sealing, Firenze, Italy, 6-8th April 1994.
6. Peppiatt, N.A. "The influence of Cylinder Tube Surface Finish on Reciprocating Seal Performance" Paper presented at 13th International Sealing Conference, 5th-6th October 2004
7. Peppiatt, N.A., "Improvements in the control of leakage from hydraulic cylinder glands". Paper presented at the 11th International Sealing Conference, Dresden, Germany, 3-4th May 1999
8. Field, G.J., "The cause and cure of interseal pressure". Fluid Power International, November 1971.

ANNEX 1 CHAPTER 15

RELEVANT NATIONAL AND INTERNATIONAL STANDARDS

STANDARD	CONTENT
BS EN 982	Safety of machinery – Safety requirements for fluid power systems and their components – hydraulics
BS EN 983	Safety of machinery – Safety requirements for fluid power systems and their components – pneumatics
ISO 1043-1	Plastics – Symbols and abbreviated terms – Basic polymers and their special characteristics
ISO 1043-2	Plastics – Symbols – Fillers and reinforcing materials
ISO 1043-3	Plastics – Symbols and abbreviated terms – Plasticizers
ISO 1629	Rubber and lattices – Nomenclature
ISO 3601-1	FP Systems – O rings – Part 1; Inside diameters, cross sections, tolerances & size identification code. (Under major revision at Feb 2007)
ISO 3601-3	FP Systems – O rings – Part 3, Quality acceptance criteria.
ISO 3601-5	FP Systems – O rings – Part 5, Materials
ISO 3939	FP Systems and components – Multiple lip packing sets – Methods for measuring stack heights.
ISO 5597	HFP – Cylinders – Housings for piston & rod seals in reciprocating applications – Dimensions & tolerances.
ISO 6072	HFP – Compatibility between elastomeric materials and fluids
ISO 6194-1	Rotary shaft lip type seals – Part 1; Nominal dimensions and tolerances
ISO 6194-2	Rotary shaft lip type seals – Part 2; Vocabulary
ISO 6194-3	Rotary shaft lip type seals – Part 3; Storage, handling and installation.
ISO 6194-4	Rotary shaft lip type seals – Part 4; Performance test procedures.
ISO 6194-5	Rotary shaft lip type seals – Part 5; Identification of visual imperfections.
ISO 6195	FP Systems and components – Single rod cylinders – Housings for rod wiper rings in reciprocating applications – Dimensions & tolerances
ISO6547	HFP – Cylinders – Piston seal housings incorporating bearing rings – Dimensions and tolerances.
ISO 6743-4	Lubricants, industrial oils and related products (class L) – Classification – Family H (Hydraulic systems).
ISO 7425-1	HFP – Housings for elastomer-energised plastic-faced seals – Dimensions and tolerances – Part 1; Piston seal housings.
ISO 7425-2	HFP – Housings for elastomer-energised plastic faced seals – Dimensions and tolerances – Part 2; Rod seal housings.

ISO 7986	HFP – Sealing devices – Standard test method to assess the performance of seals used in oil hydraulic reciprocating applications.
ISO 10766	HFP – Cylinders – Housing dimensions for rectangular-section-cut bearing rings for pistons and rods.
ISO 16589	HFP – Rotary shaft lip seals, in 5 parts.
BS 1806: 1989	Dimensions of toroidal sealing rings (O rings) & housings.
BS 4518: 1982	Specification for metric dimensions of O rings and their housings.
BS 5106: 1988	Dimensions of anti-extrusion back-up rings and their housings.
BS 7714: 2005	Guide for the care and handling of seals for FP applications.
BS 903	Physical testing of rubber.

16

POWER PACKS

Used extensively in automotive application a power pack is a combination of pump/motor reservoir and control valves to provide the power for the operation of equipment such as:

Tail gate lifts
Axle lifts
Cab lifts
Door opening
Access platforms
Scissor lifts
Small cranes / davits

Where there is a sufficient DC electrical supply a power pack can be used, and they are manufactured by specialist suppliers in very large quantities.

Large quantities mean than manufacturers can invest in high volume production of the components incorporated which has a knock on effect of lower costs which encourages more volume.

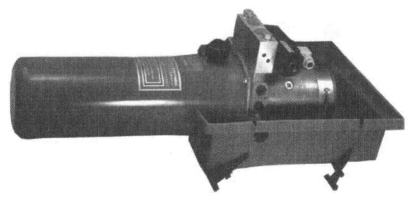

Illustration 80. DC power pack with cover removed (Fluidlink)

All the application parameters for the individual components have to be conformed with, see previous chapters. The power packs are designed for intermittent use only so heating and cooling systems are not generally required.

The speed of 'series' wound motors drops off quickly as the load is increased. 'Compound' or 'shunt' wound motors have flatter curves i.e. maintain speed at higher torques.

Brushless DC motors controlled by solid state electronics have now been developed which give a performance improvement.

High powered DC motors can be used very intermittently and the duty cycle of each application must be analysed to determine whether the motor will perform without burn out.

The curves show S2 and S3 ratings. The S2 rating = time to reach 150°C brush temperature. The S3 rating = percentage of 10 minute operating cycle.

The reservoir size is generally just enough to satisfy the displacement of the cylinders in the system.

The pump output is affected by the volumetric efficiency i.e. internal leakage and the speed of the drive motor as the load on the DC motor increases the speed drops. This depends on the type of DC motor and is illustrated by the diagram on the opposite page.

Where sustained high pressure is required i.e. work holding, clamping, compressed air driven pumps are available from specialist suppliers.

WHAT CAN GO WRONG

When filling or topping up the fluid it has to be done with all cylinders retracted otherwise the fluid returning to the reservoir will overflow when the cylinders retract.

When fitted to vehicles they need some protection from the elements and are usually provided with protective covers, as are the electrical control switches.

The pumps used are low displacement and the volumetric efficiency is important. Any damage to the pump or overheating can cause loss of pump volume and subsequent system failure.

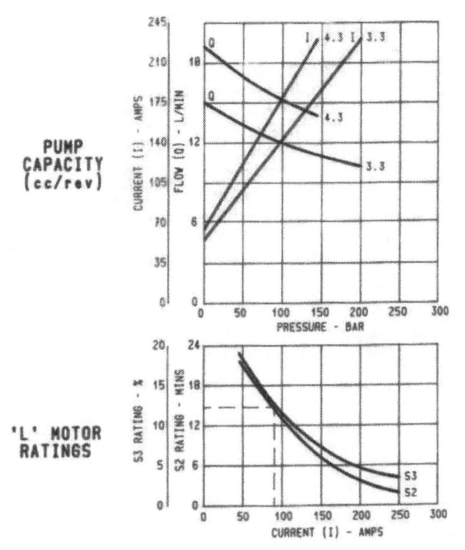

PUMP CAPACITY (cc/rev)

'L' MOTOR RATINGS

Illustration 81. Flow and current curves for DC power pack (Fluidlink)

Low DC power can also reduce pump speed and therefore volume. Battery capacity needs to match the expected duty of the power pack which can have motors up to 2 KW fitted.

More frequent use than the system is designed for can cause motor burn out.

All electrical leads, solenoid operated switches etc. need to match the capacity of the motor.

Electrical Current $= \dfrac{\text{Power (watts)}}{\text{Voltage}}$

For 2 KW motor 24 volt motor

Current $= \dfrac{2000}{24} = 83$ amps

The intermittent current can be considerably higher – see illustration 81.

17

HYDRAULIC STEERING

Many boats and vehicles incorporate power steering using hydraulic power.

Boat steering systems can be linked to an electronic compass and GPS (Global Positioning System) to provide hands off steering which when programmed with a suitable set of "way points" can literally steer a boat around the world.

Hydraulics have been used for ships' steering systems for over 100 years. The systems generally use two large low pressure cylinders in parallel operating on a quadrant attached to the rudder and have been developed by specialist steering gear manufacturers.

Vehicle steering systems i.e. cars and trucks use a combined cylinder and mechanical servo valve so that movement of the steering wheel results in displacement of the cylinder, the hydraulic supply being from an engine driven pump. Off road vehicles can have power steering which is a combination of valves, pumps and cylinders.

The steering system components are generally combinations of the pumps, valves motors and cylinders already described in previous chapters.

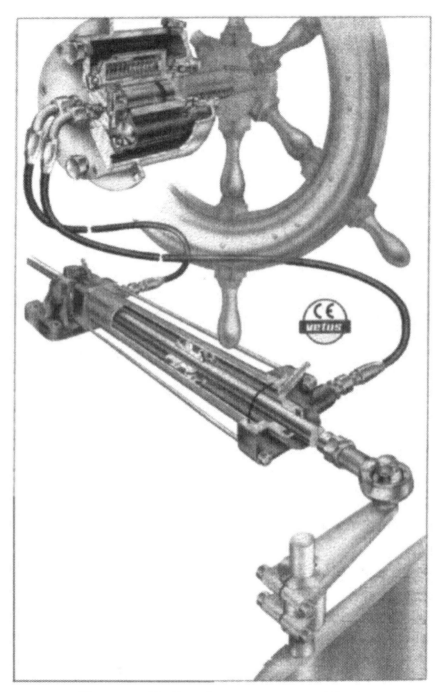

Illustration 82. Basic boat steering system (Vetus)

This illustration shows the basic system. Dual steering positions can be incorporated and autopilot introduced by adding a motor/pump unit

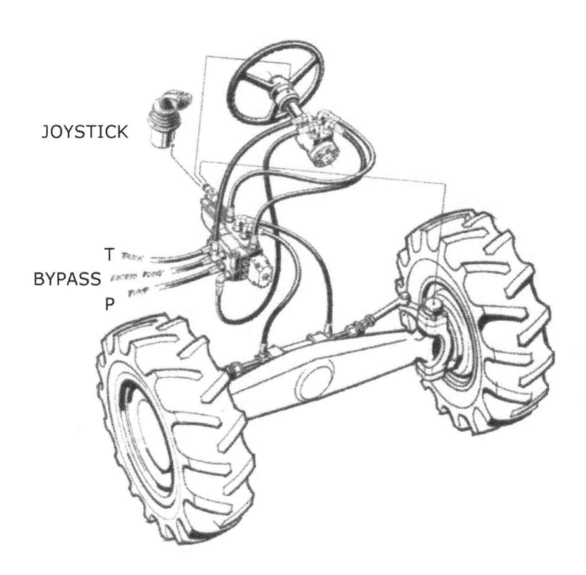

JOYSTICK

T

BYPASS

P

Illustration 83. Electro hydraulic vehicle steering (Sauer-Danfoss)

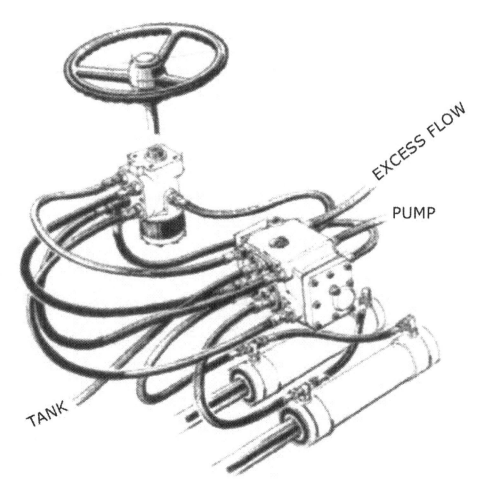

EXCESS FLOW

PUMP

TANK

Illustration 84. Hydrostatic steering for large vehicles/ships (Sauer-Danfoss)

STABILISERS

In many ways similar to a steering system but operating much faster, stabilisers incorporate a rudder like vane which protrudes from the side of the vessel and which is moved hydraulically to counter the vessel's tendency to roll. The fast, accurate response to the demands of the gyro control system requires servo or proportional valves to provide the main hydraulic control element.

WHAT CAN GO WRONG

For independent systems loss of fluid can disable the system.   Where function is critical regular checks and top-up fluid if required is essential.

Debris in the fluid can cause scoring of pump and blocking of valves.

Hydraulic hoses can be subject to mechanical damage and failure.

Ball joints can wear and cause play or backlash.

Electronic control systems on boats are liable to damage from water and vibration. Where the function is critical back up or spare electronic controls may be required.

Wiring can be undersized or corroded.

Sufficient battery power for independent systems is always a factor on yachts where long distances are concerned.

18

INSTRUMENTATION

There are three basic measurements fundamental to all fluid power systems:

Pressure
Flow
Temperature

All systems must have a pressure gauge fitted or the facility to do so to check operating pressures when commissioning the system. Mechanical gauges utilise a curved bourdon tube which opens when subject to pressure operating an indicator needle through a gear mechanism. Extremely robust these gauges can be permanently fitted. If the system is subject to spikes or shock loading they can be damped by means of a capillary tube or restrictor and many gauges are filled with glycerine to provide a damping smoothing action against both system shocks and mechanical shocks caused by vibration of the machine.

After commissioning gauges may be isolated from the system using an isolator valve to protect them and provide long life.

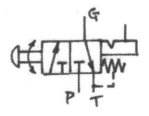

Illustration 85. Push to read gauge isolator (Isis)

Pressure can also be measured by a transducer with remote readout on the control panel. Whilst exceptionally accurate this does not provide the same sense of what is happening to the system pressure as an analogue gauge since it is virtually impossible to visually follow the changing digital output.

Flow rates can be measured by flow meters. Commonly used meters include those with a rotating element the speed of which is determined by the flow rate and which is sensed electronically and the result displayed numerically. These meters can be fitted to pressure lines as they can withstand system pressure.

Another type of meter is a moving cone in a tapered tube used in return lines. They give a visual display.

Another form is an analogue display which is operated by the flow passing a sharp edged orifice associated with a moving piston. As the flow increases the piston moves and operates the display.

Very small flow rates i.e. leakage past valves etc. are best measured by timing the flow into a cup of known volume.

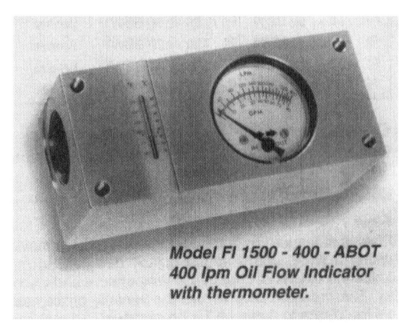

Model FI 1500 - 400 - ABOT
400 lpm Oil Flow Indicator
with thermometer.

Illustration 86. Flow meter with thermometer (HSP)

For troubleshooting commissioning purposes it is possible to obtain test units which combine both pressure and flow measurement. The result can be displayed and printed to form a permanent record.

Illustration 87. Test unit flow/pressure/temperature (Webtec)

These test units would normally be removed after the testing is complete.

A read-out of fluid temperature can often be provided from these test units.

Test points in the form of small sealable couplings can be positioned in any part of a system to enable a pressure gauge or test equipment to be plugged in.

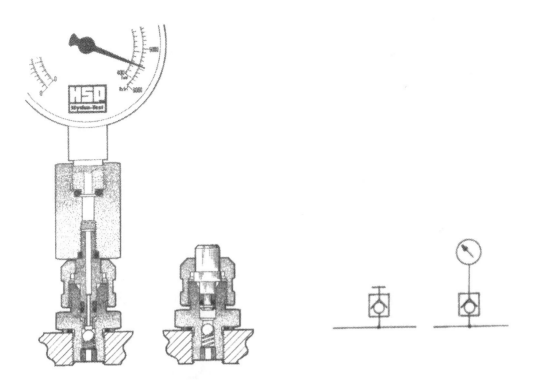

Illustration 88. Test point with gauge (HSP)

19

TEMPERATURE

A sure sign of things going wrong is a rise in temperature of the fluid. Many systems have thermometers built into the power unit often combined with the fluid level gauge.

Digital readouts combined with sensors which can be positioned practically anywhere in the system are often used.

The system operating temperature, fluid used and pump fluid viscosity requirements all need to match to obtain satisfactory performance and long life.

Systems developed in temperate climates will inevitably run hotter in tropical environments and it may be necessary to change the fluid specification i.e. viscosity to take this into account.

Although components, fluids and seals can withstand high temperatures the ideal is below 60°C (140°F). This will promote long system life and should any accident occur the fluid would not scald anyone.

The temperature range of seals may be the governing factor. At low temperatures they become hard and at high temperatures they can degrade and take up a permanent set.

See 'Seals' Chapter 15.

FLUID EXPANSION DUE TO HEAT

Mineral oil expands when heated at a higher rate than the metal tubes in which it is enclosed. The approximate fluid expansion is 1% per 14°C (30°F). If the fluid is enclosed i.e. trapped and there are large volumes of it such as in long pipelines on ships etc, a significant increase in pressure can be caused by an increase in temperature which may be sufficient to damage pipework, fitting seals etc. If the difference in temperature between night and day is 40°C (85°F) then the fluid volume would change 2.8% enough to generate over 100bar (1450psi) depending on the expansion rate of the components containing the fluid.

Pressure relieving/replenishing valves can be fitted in the pipework to eliminate this phenomenon. This is in the form of a non return valve with a direct acting preset pressure relief incorporated into it.

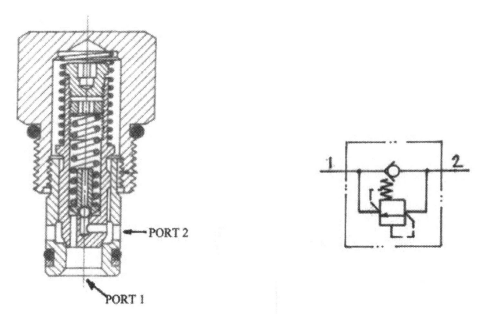

Illustration 89. Thermal relief / check in cartridge form (Eaton)

Fluid contained in long pipes which incorporate restrictors, load controls etc can get very hot if the same fluid is passed backwards and forwards over the restrictor. Avoid this by duplicating the pipework.

20

COMMISSIONING

Starting a system for the first time and after modifications, first ensure that you have hydraulic fluid of the correct grade and quality.

SIMPLE SYSTEMS

Power units, power packs which are supplied complete will have been tested by the supplier for function, leakage etc. Automotive packs will have been set up specifically for the application i.e. pressure flows etc.

Pumps can be started up off load initially by backing off the main relief valve. In many systems open centre valves will be used in the circuit which will allow flow to return to tank at low pressure.

For systems in regular production a separate flushing pump and filter can be used to clean the system to an acceptable standard before the main pumps are started.

For closed centre systems it may be a requirement to fit an air bleed valve to the output line from the pump to avoid air being trapped in that line stopping the pump from aspirating naturally. If air is trapped it can stop fluid entering the pump particularly if the pump is mounted above the level in the reservoir.

A pressure gauge must be fitted in the pump output line to ensure that the correct pressures are reached under load.

COMPLEX SYSTEMS

Have available clear hydraulic and electric circuit diagrams showing all component functions and identifying ports, solenoids etc. Also have available technical data sheets on the components used in the system.

Ensure that isolator valves in the suction and delivery pipes to the pumps are open.

Check that the pump will rotate by hand if accessible.

Pulse the motor to make the pump priming easier.

Operate at reduced pressure for a period.

It may be necessary as a precautionary measure to flush pipework systems. Subplate mounted valves can be removed and replaced with connector plates to circulate fluid around the pipework. Actuators can be bypassed for the same reason. Some systems have very long lengths of pipework particularly shipboard.

Flushing should not be the primary method of cleaning the pipework but may be considered desirable depending on the system. To be effective the flushing should be carried out under turbulent flow conditions the Reynolds Number being higher than 4,000.

Each part of the system must then be operated separately to ensure correct function.

Flow controls, pressure controls may need to be set.

Gauge points should be provided at all actuators to check that operating pressures match the design pressure.

Any system including accumulators should be commissioned with great care as accidental uncontrolled release of the contents once filled can be dangerous. Accumulator precharge pressure must be checked.

Air can get trapped in cylinders. Whilst this may become entrained in the fluid when under pressure and find its way back to the reservoir it is good practice to have the facility to bleed air from the cylinders when starting up the system as air entrained in the fluid can damage the pumps. It is not a good idea to unscrew pipe fittings to achieve this.

To assist commissioning test blocks can be inserted into pipework which measure both pressure and flow rates.

It is good practice to replace filter elements after commissioning.

When commissioning is complete, check that all system pressures are within the limits of the components used. Often pressures have to be adjusted to take into account unexpected variations in machine performance.

Never start a system unless other people are clear. If pipework or electric controls are accidentally connected up incorrectly uncontrolled operation of a machine may take place.

The author's knowledge of other people's mistakes resulting in commissioning accidents includes amputation of feet in a moulding machine, severe burns from a furnace and broken bones from an uncontrolled accumulator.

Additional reference BFPA/P3.

21

NOISE

Hydraulic pumps can be noisy. The noise originates in the pumping action and some pumps are much quieter than others. When mounted directly onto metal tanks or structures the noises can be amplified.

Generally, Internal Gear pumps are quieter than External Gear pumps and Vane pumps. Radial Piston pumps are quieter than Axial Piston pumps. Progress in noise reduction is ongoing by the pump manufacturers.

Valves can also become noisy especially as fluid viscosity is reduced when the fluid temperature increases. Most dynamic valves such as pressure controls, load controls, flow controls etc depend on orifices to provide damping to avoid instability. As the fluid gets less viscous the damping effect is reduced and if a valve becomes unstable it can be very noisy indeed.

Component manufacturers fine tune their designs to reduce noise levels but nevertheless it may be necessary to isolate pumps and also provide acoustic screening. Very effective screening is available to the marine industry in the form of sound deadening plates for engine enclosures.

To reduce pipeborne noise it may be beneficial to use a flexible pipe or some other type of barrier between the pump and the rest of the system.

For a detailed analysis of noise in hydraulic systems refer to "NOISE EMMITTED BY FLUID POWER EQUIPMENT" publication 9 by I Mech E .and BS ISO 15086 "Determination of fluid borne noise".

22

AVOIDING DISASTERS

Establish a schedule of inspection and servicing which matches the importance of system i.e. a steelworks system running continuously would be fully instrumented and monitored continuously, and maintenance is carried out when the plant is shutdown. If the system is hypercritical then back up systems will need to be available to keep the machinery running.

Inspection and maintenance of an agricultural machine may be on an as and when required basis or after the machine has been used e.g. harvester, baler.

Cure any external leaks as soon as practical.

Organise and work to a schedule for the checking and or replacement of fluid. Refer to machine supplier for guidance. Bear in mind that hydraulic oil tends to degrade much more rapidly at temperatures above 60°C [140°F].

Only use clean fluid, top up reservoirs through a filter which matches the filtration requirements of the components.

Top up using the same fluid as is already in use. Do not mix fluids without consultation with system supplier.

When replacing any component make sure it is clean inside and out.

Do not do any repairs to steel pipework, welding etc. in situ. Remove, repair, clean and refit.

If pump or component failure is frequent check out the filtration aspect of the system to ensure that it meets the requirements of the pumps and components. **A large proportion of system failures are due to contamination of the fluid.**

Also check for aeration of fluid due to leaks in suction pipes or turbulence in reservoirs.

Maintain electric systems associated with the hydraulic system. Many electrical components such as relays, microswitches etc. have far shorter life than the hydraulic components. Use protective circuits to avoid arcing at switches. Wiring can be corroded in hostile environments.

Ensure that electronic systems are not interfered with by the operation of mobile phones, security systems etc. in the vicinity.

23

SPARES

Can be from a few 'O' rings to a complete standby system depending on how critical the function of the system is.

All component manufacturers offer seal kits or spares kits containing seals, springs and other items which may need replacing.

Pump spares can include cam ring, assemblies for vane type pumps, piston assemblies for piston pumps.

Valve spares can include solenoid tubes, solenoid coils, electrical connectors, return springs etc.

Replacement hoses are a must for any system including them. Downtime due to hose failure is far more expensive than the hose itself.

A spare quantity of clean fluid of the correct specification is essential.

This book will help to establish what can go wrong and hopefully assist in establishing the maintenance and spare-parts strategy.

24

TECHNICAL DATA

Manufacturers recommended tightening torques for high tensile socket head cap screws.
For plated screws reduce tightening torque by 25% to obtain the induced load.

SIZE	TIGHTENING TORQUE		INDUCED LOAD	
	Nm	lbsft	N	lbs
M5 x 0.8	8.5	6.3	9100	2045
M6 x 1	15.5	11.4	12900	2905
M8 x 1.25	35.1	25.9	23080	5190
M10 x 1.5	69.3	51.1	36560	8220
M12 x 1.75	130	95.8	53156	11950
N°10 UNC	5.4	4	7320	1645
¼" UNC	15.0	11	13344	3000
⁵⁄₁₆" UNC	28.5	21	21530	4840
³⁄₈" UNC	58.3	43	31800	7150
⁷⁄₁₆"UNC	89.5	66	43700	9830
½" UNC	120	89	56400	12680

Important

When the body material is aluminium either a reduced torque or a longer thread engagement is required to avoid pulling the threads out of the body (minimum 2 x dia).

Where high tensile studs are used (ISO size 3 valve stacks) refer to the supplier for torque recommendations.

Fatigue

All the mechanical devices subject to fluctuating cyclic loads have a life expectancy before failure. Hydraulic components are no exception. In the USA the NFPA have a standard for fatigue testing.

Fluid power equivalents

1 bar = 10^5 N/m^2
1 bar = 10 N/Cm^2 = 1 dN/mm^2
1 pascal = 1 N/m^2
1 litre = 1000,028 cm^3
1 centistoke (cSt) = 1 mm^2/s
1 joule = 1 wattsecond (Ws)
Hertz (Hz) = cycles/second

Prefixes denoting decimal multiples or sub-multiples

For multiples

x 10^{12}	tera	T
x 10^9	giga	G
x 10^6	mega	M
x 10^3	kilo	k
x 10^2	hecto	h
x 10	deca	da

For sub-multiples

x 10^{-1}	deci	d
x 10^{-2}	centi	c
x 10^{-3}	milli	m
x 10^{-6}	micro	μ
x 10^{-9}	nano	n
x 10^{-12}	pico	p
x 10^{-15}	femto	f
x 10^{-18}	atto	a

Conversion factors

To convert → Into →		Into → To convert →		Multiply by / Divide by
Unit	Symbol	Unit	Symbol	Factor
BTU/hour	Btu/h	kilowatts	kW	0.293071×10^{-3}
Cubic centimetres	cm^3	litres	l	0.001
Cubic centimetres	cm^3	millilitres	ml	1.0
Cubic feet	ft^3	cubic metres	m^3	0.0283168
Cubic feet	ft^3	litres	l	28.3161
Cubic inches	In^3	cubic centimetres	cm^3	16.3871
Cubic inches	In^3	litres	l	0.0163866
US galls	USg	litres	l	3.785
Degrees (angle)	O	radians	rad	0.0174533
Fahrenheit	°F	Celsius (centigrade)	°C	* see below
Feet	ft	metres	m	0.3048
Feet of water	ft $H_2$0	bar	bar	0.0298907
Foot pounds f.	ft lbf	joules	J	1.35582
Foot pounds/minute	ft lbf/min	watts	W	81.3492
Gallons, UK	UK gal	litres	l	4.54596
Gallons, US	US gal	litres	l	3.78531
Horsepower	hp	kilowatts	kW	0.7457
Inches	in	centimetres	cm	2.54
Inches	in	millimetres	mm	25.4
Kilogramme force	kgf	newtons	N	9.80665
Kilopascals	kPa	bar	bar	0.01
Kiloponds	kp	newtons	N	9.80665
Kilopond metres	kp m	newton metres	Nm	9.80665
Kiloponds/square centimetre	kp/cm^2	bar	bar	0.980665
Metric horsepower		kilowatts	kW	0.735499
Microinches	μin	microns	μm	0.0254
Newtons/square centimetre	N/cm^2	bar	bar	0.1
Newtons/square metre	N/m^2	bar	bar	10^{-5}
Pascals (newtons/sq metre)	Pa	bar	bar	10^{-5}
Pounds (mass)	lb	kilogrammes	kg	0.4536
Pounds/cubic foot	lb/ft^3	kilogrammes/cubic metre	kg/m^3	16.0185
Pounds/cubic inch	lb/in^3	kilogrammes/cubic centimetre	kg/cm^3	0.0276799
Pounds force	lbf	newtons	N	4.44822
Pounds f. feet	lbf ft	newton metres	Nm	1.35582
Pounds f. inches	lbf in	newton metres	Nm	0.112985
Pounds f./square inch	lbf/in^2	bar	bar	0.06894
Revolutions/minute	r/min	radians/second	rad/s	0.104720
Square feet	ft^2	square metres	m^2	0.092903
Square inches	in^2	square metres	m^2	6.4516×10^{-4}
Square inches	in^2	square centimetres	cm^2	6.4516

* $°C = 5(°F - 32)/9$

Practical hydraulic formulae

Geometric flow rate (l/min)
(pumps and motors)
$$= \frac{\text{Geometric displacement (cm}^3/\text{r) x shaft speed (r/min)}}{1000}$$

Theoretical shaft torque (Nm)
(pumps and motors)
$$= \frac{\text{Geometric displacement (cm}^3/\text{r) x pressure (bar)}}{20\pi}$$

Shaft power (kW)
$$= \frac{\text{Torque at shaft (Nm) x shaft speed (r/min)}}{9550}$$

Hydraulic power (kW)
$$= \frac{\text{Flow rate (l/min) x pressure (bar)}}{600}$$

Heat equivalent of hydraulic power (kJ/min) $= \dfrac{\text{Flow rate (l/min) x pressure (bar)}}{10}$

Geometric flow rate (l/min)
(cylinders)
$$= \frac{\text{Effective area (cm}^2) \text{ x piston speed (m/min)}}{10}$$

Theoretical force (N)
(cylinders)
$$= \text{Effective area (cm}^2) \text{ x pressure (bar) x 10}$$

Velocity of fluid in pipe (m/s)
$$= \frac{\text{Flow rate (l/min) x 21.22}}{D^2}$$
where D = inside diameter of pipe in millimetres

25

INDUSTRIAL STANDARDS/GUIDELINES

NATIONAL & INTERNATIONAL STANDARDS & CETOP RECOMMENDATIONS

BSI and ISO standards are available for reference in most large reference libraries and are available for purchase from BSI Sales, Customer Services, 389 Chiswick High Road, London W4 4AL Tel: 0181 996 7000, Fax: 0181 996 7001. CETOP (European Oil Hydraulic and Pneumatic Committee) recommendations are available from BFPA.

The following lists of standards and recommendations are selected from the hundreds of fluid power standards which exist, as having the most relevance to design aspects of the fluid power categories in which they are listed.

The designer needs to be aware of CEN standards associated with the Machinery Directive, 98/37/EEC, etc., such as EN292, EN414, EN982 and EN983.

Many of the British Standards are similar or identical to some of the ISO Standards.
No equivalence is implied by reading across the tables.

Subject	International Standards (ISO) British Standards (BS) European Standards (EN)
Accumulators	
Characteristics	BS ISO 5596
Gas ports	BS ISO 10945
Hydraulic ports	BS ISO 10946
Design	BS EN 14359
Hydraulic Pumps, Motors & Integral Transmissions	
Flanges - Dimensions	ISO 3019 BS 6276
Geometric displacements	ISO 3662
Parameter definitions and letter symbols	ISO 4391
Motor characteristics	BS ISO 4392
Motor characteristics - constant flow	ISO 4392-3 BS 7275-3
Steady state performance	ISO 4409 BS 4617
Pumps/Motors - airborne noise	ISO 4412 BS 5944
Positive displacement - derived capacity	ISO 8426 BS 5944
Pumps - Pressure ripple levels	BS ISO 10767
Cylinders	
Bores/piston rod diameters	BS ISO 3320 & 3321
Nominal pressures	BS ISO 3322
Piston strokes - basic series	BS ISO 4393
Cylinder barrels — requirements	ISO 4394-1 BS 5242-1
Piston rod dimensions	BS ISO 4395
Piston rod housings — dimensions	ISO 5597 BS 5751
Mounting dimensions	BS ISO 6020
Mounting dimensions - single rod 250 bar series (hyd.)	BS ISO 6022
Mounting dimensions - identification code	BS ISO 6099
Mounting dimensions - single rod 10 bar series (pneu.)	ISO 6430, 6431, 6432
Cylinder barrels - requirements (pneu.)	ISO 6537 BS 5242-2
Mounting dimensions - rod end (hyd.)	BS ISO 6981, 6982
Bore and port thread sizes (pneu.)	BS ISO 7180
Bore and rod area ratios (hyd.)	BS ISO 7181
Single rod (hyd.)	BS ISO 8131, 8132, 8133, 8135, 8137, 8138
10 bar series - mounting dimensions (pneu.)	BS ISO 8139, 8140
Pneumatic cylinders - acceptance test	BS ISO 10099
Hydraulic cylinders - acceptance test	BS ISO 10100
Mounting dimensions - 100 bar series (hyd.)	BS ISO 10762
Method for determining buckling load	ISO/TS 13725
Single rod, 160 bar compact series - accessory dimensions (hyd.)	BS ISO 13726
Pneumatic slides — load capacity and presentation method	ISO/TR 16806
Plastics hoses - textile reinforced compressed air type	BS EN ISO 5774
Pipes & Couplings	
Pipe threads - pressure tight joints not made on threads	BS EN ISO 228
Nominal tube OD and hose ID	ISO 4397
Nominal pressures	ISO 4399 BS 7847
Ports & stud ends with ISO 261 threads and O-ring sealing	BS ISO 6149
Pneumatic cylindrical quick-action couplings	ISO 6150
Hydraulic flange, split or one-piece, metric or inch	BS ISO 6162
Hydraulic flange - four-screw, 250 bar and 400 bar	ISO 6164 BS 7504 BS 7504
Hydraulic quick-action couplings	ISO 7241 BS 7198-1
Metallic tube connections	BS EN ISO 8434
Metallic tube connections	ISO 8434 BS 4368-4 & BS DD 195 (not equivalent)
Ports and stud ends with ISO 261 threads	BS EN ISO 9974
Plain-end, seamless and welded precision steel tubes	ISO 10763
Ports and stud ends with ISO 725 threads and O-ring sealing	ISO 11926
Hose fittings	BS ISO 12151
Hydraulic couplings for diagnostic purposes	BS ISO 15171
Hydraulic flush-face type quick-action couplings	BS ISO 16028
Pneumatic connections - port and stud ends	BS ISO 16030

Hydraulic Hose & Hose Assemblies	
Rubber and plastics hoses and hose assemblies - testing	BS EN ISO 1402
Rubber hoses - wire-reinforced (non-water based fluids)	ISO 1436-1 BS EN 853
Rubber hoses - bending tests	BS EN ISO 1746
Rubber hoses - spiral wire-reinforced (non-water based fluids)	ISO 3862-1 BS EN 856
Thermoplastics hoses - textile reinforced	ISO 3949 BS EN 855
Rubber hoses - textile reinforced (non-water based fluids)	ISO 4079-1 BS EN 854
Rubber hoses - measurement methods	BS EN ISO 4671
Rubber hoses - determination of abrasion resistance	BS EN ISO 6945
Rubber hoses - wire reinforced 'compact' (non-water based fluids)	ISO 11237-1 BS EN 857
Hose assemblies - method of test	BS ISO 6605
Hydraulic hose assemblies - external leakage classification	ISO/TR 11340
Control Components	
Three-pin electrical plug connectors with earth contact	ISO 4400 BS 6361
Hydraulic four-port directional control valves	BS ISO 4401
Hydraulic valves - pressure differential/flow characteristics	IS0 4411 BS 4062-1
Pneumatic five-port directional control valves	BS ISO 5599
Hydraulic pressure valves	BS ISO 5781
Hydraulic valve mounting surfaces and cartridge valve cavities	ISO 5783
Compensated flow-control valves (hyd.)	BS ISO 6263
Hydraulic pressure-relief valves	BS ISO 6264
Hydraulic valves controlling flow and pressure	ISO 6403 BS 4062-2
Two-pin electrical plug connectors with earth contact	ISO 6952
Hydraulic two-port slip-in cartridge valves	ISO 7368 BS 7296-1
Hydraulic two-, three- and four-port screw-in cartridge valves	BS ISO 7789
Hydraulic four-port stack valves and directional valves	BS ISO 7790
Hydraulic valve ports, subplates, control devices and solenoids	9461
Hydraulic four- and five-port servovalves	BS ISO 10372
Hydraulic electrically modulated directional flow valves	BS ISO 10770
Ports and control mechanisms of pneumatic valves and components	BS ISO 11727
Pneumatic directional valves - shifting time	BS ISO 12238
16mm square electrical connector with earth contact	BS ISO 15217
Pneumatic fluid power. 3/2 solenoid valves	BS ISO 15218
Pneumatic five-port directional control valves	BS ISO 15407
Pressure switches - mounting surfaces (hyd.)	BS ISO 16873
Pneumatic FRLs	
Compressed-air filters - characteristics and test	BS ISO 5782
Compressed-air lubricators - characteristics and test	BS ISO 6301
Pneumatic filter-regulators - characteristics and test	BS ISO 6953
Compressed air - contaminants, purity classes and particle content	BS ISO 8573
Contamination control – Hydraulic	
Hydraulic filter elements - collapse/burst resistance	BS ISO 2941
Hydraulic filter elements - fabrication integrity, determination of first bubble point	BS ISO 2942
Hydraulic filter elements - material compatibility with fluids	BS ISO 2943
Hydraulic fluid sample containers	BS ISO 3722
Hydraulic filter elements – end load test method	BS ISO 3723
Hydraulic filter elements – flow fatigue characteristics	BS ISO 3724
Contamination analysis - method for reporting data	ISO 3938
Hydraulic filters - pressure drop versus flow characteristics	BS ISO 3968
Particulate contamination analysis - extraction of fluid samples	ISO 4021 BS 5540-3
Determination of particulate contamination, gravimetric method	ISO 4405 BS 5540-7
Hydraulic fluids - method for coding the level of contamination	BS ISO 4406
Hydraulic fluid contamination - particulate contamination	BS ISO 4407 and BS ISO 11500
Guidelines for controlling component cleanliness	ISO/TR 10949
Hydraulic filter elements - performance characteristics	BS ISO 11170
Calibration of automatic particle counters for liquids	BS ISO 11171 BS ISO 11943
Calibration of liquid automatic particle counters	ISO/TR 16144
Impact of changes in ISO fluid power particle counting	BS ISO/TR 16386
Assembled hydraulic systems - verification of cleanliness	ISO/TS 16431
Hydraulic filters - performance, multi-pass test method	BS ISO 16889
Cleanliness of parts and components - inspection and data reporting	BS ISO 18413

Seals and their housings	
O-rings - dimensions and quality	ISO 3601 BS 6442 ISO updated recently
Multiple lip packing sets - methods for measuring stack heights	ISO 3939
Compatibility between hydraulic fluids and standard elastomeric materials	BS ISO 6072
Rotary shaft lip type seals	ISO 6194 BS 7780
Housings - rod wiper rings — dimensions	BS ISO 6195
Piston seal housings -dimensions (hyd.)	BS ISO 6547
Dimensions - seal housings (hyd.)	BS ISO 7425
Sealing devices - standard test methods for reciprocating applications	ISO 7986
Rectangular bearing rings for piston/rods - housing dimensions (hyd.)	BS ISO 10766
Rotary shaft lip-type seals incorporating thermoplastic sealing elements	BS ISO 16589
Hydraulic Fluids	
Classification of family H - hydraulic fluids	BS EN ISO 6743-4
Fire resistant fluids - guidelines for use	ISO 7745
Mineral oil specifications	BS ISO 11158
Fire-resistant fluid specifications	BS EN ISO 12922
Eco-friendly fluid specifications	BS ISO 15380
Filterability of fluids - test method	BS ISO 13357
Fire resistant fluids - guidelines for selection	BS EN 14489
Fire resistant fluids - wick test	BS EN ISO 14935
Fire resistant fluids - spray test	BS EN ISO 15029-1
Systems & General	
Safety of machine tools - Hydraulic presses	BS EN 693
Safety of machinery - Hydraulic systems and their components	BS EN 982
Safety of machinery - Pneumatic systems and their components	BS EN 983
Symbols	BS ISO 1219-1
Circuit diagrams	BS ISO 1219-2
Nominal pressures	BS ISO 2944
Hydraulic systems - general rules	BS ISO 4413 EN 982 (not equivalent)
Pneumatic systems - general rules	BS ISO 4414 EN 983 (not equivalent)
Fluid power vocabulary — bilingual	ISO 5598
Fluid logic circuits — symbols	ISO 5784
Flow-rate characteristics for pneumatic components	ISO 6358 BS 7294
Hydraulic gantry lifting gear	BS 7121-13
Pneumatic standard reference atmosphere	BS ISO 8778
Measurement techniques (hyd.)	ISO 9110
Fatigue pressure testing of metal pressure-containing envelopes (hyd.)	BS ISO 10771-1
Safety of machine tools - Hydraulic press brakes	BS EN 12622
Determination of fluid borne noise	BS ISO 15086

A Selection of Guideline Documents

BFPA/P3	1995	Guidelines for the Safe Application of Hydraulic and Pneumatic Fluid Power Equipment	**BFPA/P53**	2002	Fluid Power at the Forefront
BFPA/P4	1986	Guidelines for the Design of Quieter Hydraulic Fluid Power Systems (Third Edition)	**BFPA/P54**	2003	Guidelines to the Pressure System Safety Regulations 2000 and their Application to Gas-loaded Accumulators
BFPA/P5	1999	Guidelines to Contamination Control in Hydraulic Fluid Power Systems	**BFPA/P55**	1993	Guidelines for the Comparison of Particle Counters and Counting Systems for the Assessment of Solid Particles in Liquid
BFPA/P7	2004	Guidelines to the Design, Installation and Commissioning of Piped Systems Part 1 - Hydraulics	**BFPA/P56**	2004	BFPA Fluid Power, Engineer's Data Booklet
BFPA/P9	1992	Guidelines for the Flushing of Hydraulic Systems	**BFPA/P57**	1993	Guidelines to the Use of Ecologically Acceptable Hydraulic Fluids in Hydraulic Fluid Power Systems
BFPA/P12	1995	Hydraulic Fluids Mineral Oil Data Sheets	**BFPA/P58**	2003	The Making of Fluid Power Standards
BFPA/P13	1996	Fire-Resistant Hydraulic Fluids Data Sheets	**BFPA/P59**	1993	Proceedings of the 1993 BFPA Leak Free Hydraulics Seminar
BFPA/P22	2003	Guidelines on Selection of Industrial O- Rings (Metric & Inch)	**BFPA/P60**	1994	Leak-Free for Hydraulic Connections
BFPA/P27	1993	Guidelines on Understanding the Electrical Characteristics of Solenoids for Fluid Power Control Valves & their Application in Potentially Explosive Atmospheres	**BFPA/P61**	1998	A Guide to the Use of CE Mark
BFPA/P28	1994	Guidelines for Errors and Accuracy of Measurements in the Testing of Hydraulic & Pneumatic Fluid Power Components	**BFPA/P65**	1995	VDMA 24 568 & 24 569 Rapidly Biologically Degradable Hydraulic Fluids Minimum Technical Requirements & Conversion from Fluids based on Mineral Oils
BFPA/P29	1987	General conditions for the Preparation of Terms and Conditions of Sale of UK Fluid Power Equipment Manufacturers and Suppliers	**BFPA/P66**	1995	BFPA Survey on Ecologically Acceptable Hydraulic Fluids
BFPA/P41	1995	Guidelines to Hydraulic Fluid Power Control Components	**BFPA/P67**	1996	Ecologically Acceptable Hydraulic Fluids Data Sheets
BFPA/P44	1995	Index of BS/ISO Standards Relating to Fluid Power	**BFPA/P68**	1995	Machinery Directive Manufacturers Declarations
BFPA/P47	2004	Guidelines to the Use of Hydraulic Fluid Power Hose and Hose Assemblies	**BFPA/P83**	2003	The World of Fluid Power 2003 CD Rom (Edition 3)
BFPA/P48	1998	Guidelines to the Cleanliness of Hydraulic Fluid Power Components	**BFPA/P95**	2003	Principles of Hydraulic System Design
BFPA/P49	1995	Guidelines to Electrohydraulic Control Systems	**BFPA/P100**	2003	Guidelines for the Proof & Burst Pressure Testing of Fluid Power Components
BFPA/P52	1997	Guidelines to the Plugging of Hydraulic Manifolds and Components	**BFPDA/D2**	1994	Technical Guidelines for Distributors of Hydraulic Fluid Power Equipment

INDEX

Page numbers in **bold type** indicate main references.

Lightning Source UK Ltd.
Milton Keynes UK
UKOW07f0122050615

252937UK00001B/45/P